P9-EDF-350

STOP, THIEF!

INFORMATION RESOURCES CENTER
ASIS
1625 PRINCE STREET
ALEXANDRIA, VA 22314
tel. (703) 519-6200
8006
MAR 3 1 1993

TH
9739
.M465
1978

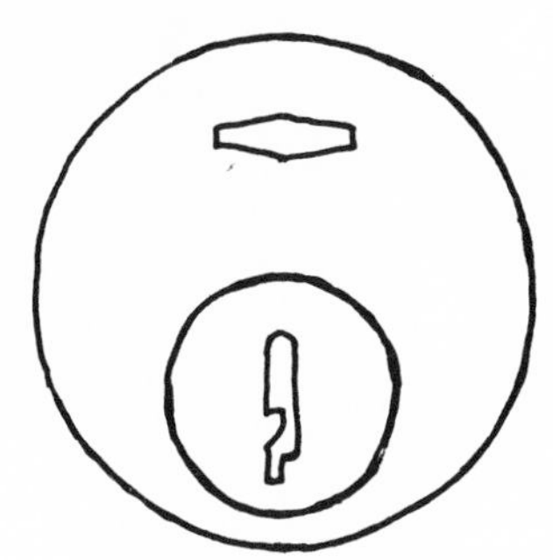

How STOP, THIEF! to safeguard & secure your home & business

ROBERT McDERMOTT

THEODORE IRWIN

Macmillan Publishing Co., Inc. NEW YORK

Copyright © 1978 by Robert McDermott and Theodore Irwin

All rights reserved. No part of this book may be reproduced or transmitted in any form or by any means, electronic or mechanical, including photocopying, recording or by any information storage and retrieval system, without permission in writing from the Publisher.

Macmillan Publishing Co., Inc.
866 Third Avenue, New York, N.Y. 10022
Collier Macmillan Canada, Ltd.

Library of Congress Cataloging in Publication Data

McDermott, Robert.
Stop, thief!

Includes index.
1. Burglary protection. 2. Security systems.
I. Irwin, Theodore D., date joint author.
II. Title. III. Title: How to safeguard & secure your home & business.
TH9705.M33 643 77-28690
ISBN 0-02-583080-5

First Printing 1978

Designed by Jack Meserole

Printed in the United States of America

Material from *Crime and the Mind* by Walter Bromberg, M.D. Copyright 1948 by J. B. Lippincott Company. Reprinted by permission.

AUTHORS' NOTE

Certain names and locations in this book have been changed to protect the innocent or to avoid embarrassment and undue publicity.

To HELEN MCDERMOTT *and* HELEN IRWIN

for their abiding fortitude

Contents

Acknowledgments

The authors acknowledge with appreciation the cooperation of Mort Weisinger and Detectives John Kid, Joseph Kiernan, and Martin Gillen in providing us with their experiences; and express our gratitude to Kathy Kainer for her generous help in preparation of the manuscript.

Foreword

Thieves respect property. They merely wish the property to become their property so that they may more perfectly respect it.
—CHESTERTON

EVERY TEN SECONDS a burglary is committed in the United States.

It happens in suburban homes and city apartments, in offices and stores, in banks and warehouses. Two out of three burglaries strike at residences. You—or a neighbor, friend, or relative—may have been a victim or will be before this year is out.

The toll of this criminal plague: $1.4 billion a year. And the FBI Uniform Crime Reports reveal that the incidence of burglary offenses has soared 47 percent in the past five years.

The burglary toll goes far beyond the average dollar loss of $422 per victim. Burglary—that is, surreptitious or forced entry—often becomes robbery. It may lead to assault, rape, or murder. It was the bungled burglary at Watergate that triggered a national political upheaval, changing the course of history. An earlier successful entry led to another shocking crime of the century, the Lindbergh kidnapping and murder.

Burglary waves have terrorized people. In one suburb, a family's garden apartment was plundered twice within six months. Gone were the color TV set, hi-fi, cameras, golf clubs, best clothes, and jewelry. "We'll just have to move," the family decided. "It's the only way not to be ripped off again."

It's *not* the only way. With sensible preventive and defensive measures against uninvited larcenous visitors you can avoid becoming a sitting duck for break-in artists; you *can* frustrate and ward them off.

This is what *Stop, Thief!* is all about. Regard it as your manual of crime deterrence, a practical do-it-yourself guidebook to safe-guarding your property, your family, your life.

Our aim is to make the thief's life more difficult and your life easier. It is hoped that this book will provide a sense of security to home dwellers and apartment tenants, as well as to business and professional people. Housewives, the elderly, and people living alone will especially find it a source of reassurance.

You will learn how and where your dwelling or commercial premises may be vulnerable. By understanding how the amateur or professional burglar operates, you can outwit him. If he does somehow break in, you'll know how to save your life.

Just who are your antagonists? They come in two breeds: the crude amateurs and the professional career burglars. Most of the crude type are narcotic addicts or teenagers, working in less affluent neighborhoods. Police have compared them to rats: they lack courage but if cornered they may turn bold and desperate, and attack occupants of a home or office.

The skilled professional burglar goes for the big "score" in better residences, hotels, and business establishments. He seeks jewelry, furs, and cash in residences; travelers' checks, money orders, and cash in business safes. He dresses well, plans his jobs carefully, and constantly sharpens his talents.

Let's see what makes a professional burglar tick. When behavioral psychologists and psychiatric criminologists theorize about him they come up with assorted "neurotic mechanisms," but these may not really explain every situation. Consider some of their concepts and assumptions.

A distinction should be made between burglary and robbery. A *burglar* enters a dwelling or other premises with the intent to steal, usually in the absence of the owner or in the belief that no one is there. A *robber*, having entered, applies direct force or violence on a person he believes to have money or other valuables or who tries to obstruct his mission.

Burglary thus becomes, presumably, a stealthy, passive crime; robbery, an aggressive breach of the law. Yet psychiatrists tell us that stealth is really aggression under cover, the assailing impulses of a seemingly passive individual haunted by a feeling of inferiority. The choice of burglary as a mode of larceny implies furtiveness in the criminal's makeup. Psychologists see this as suppressed hostility toward a parent or parent surrogate and it represents "a reaction to infantile feelings of deprivation—unquestionably neurotic in nature."

Focusing on chronic burglars, Dr. Walter Bromberg, former director of the Psychiatric Clinic at New York's Court of General Sessions, once analyzed them this way in his book, *Crime and the Mind*:

"Professional burglars embark on a career of crime at a time of economic necessity and continue in it for the profit involved. They are men of mature years, accepting the advantages and risks of this type of work as one accepts those of a legitimate profession. Never outspokenly aggressive, even when apprehended, they express themselves in subverted behavior. The explanation usually given—that their criminal careers were due to circumstance—is a rationalization. Acceptance of a criminal pattern of life by chronic offenders provides them, in actuality, with a feeling of security. They sink into the anonymity of burglary to protect unperceived neurotic impulses."

The psychiatrist went on to say that the "compulsive" quality of burglary drives thieves to continue for years without involving a single instance of frank aggression, such as robbery. "The chronic burglar," he concludes, "who has consolidated his neurotic impulses into an anti-social personality is in a fact a psychopathic individual."

Behavioral psychologists and psychiatrists fall into the trap of lumping *all* burglars into a stereotype. They tend to view a burglar as a lone individual tiptoeing surreptitiously through a darkened residence, a small flashlight in his gloved hand, stealing trinkets and masturbating into a bedsheet. Although unquestionably this fellow exists, what of the many others in the trade who are totally unlike him?

The burglars who once used a sledgehammer to smash Tiffany's showcase window on Fifth Avenue in New York at 2:00 A.M. certainly were not secretive. Nor was stealth evident when burglars used the chain from a tow truck to wrap around a 450-pound safe and pull it out of a second-story window to the street below, incidentally tearing a five-foot hole out of the side of the building.

In fact, it is *boldness*, not furtiveness, that is characteristic of most successful burglars. With singular poise they can operate in daylight hours, dressed in shirt, tie, and conventional business suit. Almost always they are glib con men, able to talk their way out of sticky situations when encountered where they shouldn't be.

Take, for example, the consummate coolness of that veteran craftsman, Seymour Refkin, a middle-aged, smartly tailored character who might be mistaken on the street for a thriving lawyer or stockbroker. One afternoon while plying his trade in a midtown luxury apartment house Seymour was surprised to hear someone rattling keys at the door of the apartment where he had picked the lock. Having neglected to plug the cylinder (to prevent anyone from opening the lock), Seymour whipped off his hat and overcoat, picked up a magazine, and relaxed on a sofa, nervous but thankful that he had noted the name of a doctor on the floor below. The tenant of the apartment in which Seymour now sat started to enter—and stood paralyzed in the doorway.

"What are you doing here?" he gasped.

Pretending to be startled, Seymour graciously replied, "Why, I'm waiting to see the doctor." And the burglar nonchalantly kept reading his magazine.

At the threshold the tenant, a rugged young man, stared at the number on his door to be sure *he* wasn't in the wrong apartment.

"Doctor? What doctor?" the man demanded.

"Doctor Goldstern, apartment Eighteen E."

"This is Nineteen E!"

"Oh? You sure? I'm sorry, fella." And Seymour, scratching his head in puzzlement, picked up his hat and coat and moved to leave.

The tenant asked, "By the way, how did you get in here?"

Seymour shrugged. "Doctor Goldstern told me to come right up. I rang the bell, got no answer, and figured he was with another patient. I tried the door, it was unlocked, and I just walked in. Just been here a few minutes."

With unassailable aplomb the burglar sauntered out—the loot in his pockets.

Smooth and slippery, burglars like Seymour, fairly typical of the professional breed, are not exactly what psychologists envision. Unless they can equate "stealthy" with "Machiavellian."

As for motivation, the professional is mindful of security as well as economics in his choice of a criminal calling. The skilled burglar reaps much greater monetary rewards than does the holdup man; and chances of the burglar avoiding being killed or arrested during a performance are much better than those of the armed heistman. Dr. Bromberg contended that burglars embark on a career in crime "at a time of economic necessity" and stay in it for the profit. But he later contradicted himself by observing that "this is a rationalization." Whatever the interpretation, or convoluted jargon used, safety—as well as lucre—may be more intrinsic to the intelligent thief's penchant for burglary than any perception of inferiority. Far from feeling inadequate, he tends to regard himself as virtually omnipotent.

"There's no place I can't get in," a career burglar once boasted in a masked television interview. "No lock I can't take, no safe, no alarm I can't beat." But the following week an effective alarm system landed him in jail.

Stop, Thief! is the product of Robert James McDermott's experiences as the burglar's bête noire in New York, the lock-picker's paradise. During the course of two decades until his retirement as Detective First Grade in the Safe, Loft, and Burglary Squad he was instrumental in putting hundreds of break-in practitioners behind bars.

Acknowledging Mac's unique talents, the Police Department officially designated him as its burglary expert. Teletypes issued to all detective commands directed them to notify McDermott in all cases of unusual burglaries or other crimes where the method of entry or modus operandi seemed mystifying. His testimony as an expert witness—in cases ranging from simple theft to grisly homicide—was and still is accepted in court as gospel from The Master. Ineluctably the detective's reputation spread nationwide and abroad.

"Burglary," he maintains, "is a crime of opportunity. The victim contributes to his loss by a lack of awareness of the inherent weaknesses in security devices he depends on. I feel the public, concerned and fearful, is anxious for competent advice."

In scores of lectures at universities, police organizations, and insurance gatherings, McDermott's shrewd insights into the burglarious mind and psyche have proved enlightening even to veteran cops. Again and again, spellbound fellow law enforcers have prodded him: "Mac, why don't you write a book?"

Stop Thief! is the book. With it you should be able to leave your home, or go to sleep, with a comforting sense of freedom from fear of burglars.

Note: In the course of this book it may appear that we are telling thieves how to commit burglary. We believe that an *informed public* is the most effective answer to the zooming burglary statistics. We are convinced that when you—the victim or potential victim—are aware how burglars operate, you will better be able to outwit them. When consumers can recognize the flaws in common locks, alarms, and similar defensive devices, and demand more effective equipment, manufacturers may be more willing to research, develop, and make available better security products.

STOP, THIEF!

Chapter 1 How you become a victim: The burglar's ruses and stratagems

IT HAPPENED recently at a twenty-one-story high-rental apartment house in an upper-middle-class neighborhood. The FBI's New York headquarters was just up the street, a police station two blocks away. A maid reporting for work in the morning entered the seventh-floor, five-room apartment of Mr. and Mrs. Lawrence Gerber and stood frozen in the foyer.

Mr. Gerber, an eighty-four-year-old lawyer, lay sprawled on the floor, strangled with a necktie. His face was bruised, as if he had been struck by a fist or some hard object. In the master bedroom frail Mrs. Gerber, seventy-six, who had been confined to a wheelchair, had also been strangled.

The apartment had been thoroughly ransacked. Everything was in disarray. The contents of dressers and chests had been dumped onto a bed as if in a frantic search. Clothes in the closet were askew.

Police were summoned. A medical examiner estimated the old couple had been dead at least twelve hours when they were discovered. Detectives found no sign of forced entry on the apartment door. The lock had not been picked.

After questioning the 250 other tenants in the building, police revealed there had been nineteen break-ins over the previous three years. In a dozen cases, keys had been used or locks picked. On three occasions locks had been forced. The remaining four thefts were accomplished by intruders who probably used a ruse to gain entrance.

The building had a doorman who ostensibly announced and checked each visitor. There were times, though, when he was out hailing cabs for tenants, so intruders could easily walk in and out unquestioned. The rear service entrance was often left open, not locked until 7:00 P.M. The garage entrance was open until 1:00 A.M. Both of these entrances led to the stairways and elevators inside the building. Either would have enabled the slayer—any burglar—to slip into the house.

The Gerber entry had not been forced. Had the killer been admitted voluntarily? If so, what stratagem had he used?

One clue came from a fifteenth-floor tenant, Robert Bobson, a retired bank vice-president. Seven months before, his apartment had been burglarized in his absence. About $71,000 worth of jewelry was stolen. Police had found no sign of forced entry. Now Mr. Bobson recalled that on the afternoon of the day the Gerbers were murdered, he was told by his housekeeper that an insurance man was on his way up to see him. But Mr. Bobson had made no such appointment.

"We locked the door," he related, "and when the man arrived I would not open it. He said he was from Aetna and that I had asked about buying a policy. I said, 'No,' then turned to my housekeeper and in a loud voice told her to call the police."

The "insurance" man quickly left. Presumably the trick had worked with the naive Gerbers.

A week later in Scarsdale, a New York suburb, Mrs. Phoebe Langley had lunch with a friend at a local restaurant. Leaving, she missed her purse. It had, she thought, been placed on the floor beside her chair. No one in the restaurant could find it. Upset, uncertain what to do about it, Mrs. Langley returned home.

Two hours later the phone rang. It was a woman's voice, apparently cultured.

"Hello. May I speak with Mrs. Phoebe Langley? Mrs. Langley? My name is Jane Foster and I have your purse. . . . I found it in the parking lot outside a Scarsdale restaurant. I'd just had lunch with a friend and went to my car and there was your purse, as if it had been thrown on the hood. I had to drive my friend back to Manhattan and that's where I am now, in her apartment. We found your name in the purse and my friend got your phone number from Information. . . . Oh, Mrs. Langley, there's no money at all in your wallet but you'll be glad to know your credit cards and keys are safe. Tell me, what happened?"

Phoebe Langley related her experience in the restaurant and told "Jane" she'd had about sixty dollars in her wallet. The woman assured her the purse would be returned. Could they meet at the same restaurant—for lunch? Tomorrow? Much relieved, Mrs. Langley agreed.

Next day she met a well-dressed matronly woman at the restaurant and gratefully received her purse. They had a few drinks, chatted, and enjoyed a leisurely lunch.

At home, an hour and a half later, Mrs. Langley had the shock of her life. The house had been stripped of everything valuable that could be taken away: her fur coat, a six-thousand-dollar painting, her husband's suits, a typewriter, stereo equipment, even some kitchen appliances. A panel truck, she learned later, had driven up, a man casually opened the door with a key and carted out the loot.

The purse thief and her partner had duplicated the house key in the purse; an I.D. card and letters had provided the address.

Let's observe Phil Rosen, a sandy-haired, blue-eyed young itinerant photographer. Driving through a Philadelphia suburb he swore that if he ever got married they'd raise beagles, not babies. Taking pictures of brats had been extraordinarily profitable lately; he'd made three really good "scores" in the past month.

Today his notebook bore this notation: "Anderson to Scranton all day." Mrs. Thomas Anderson was typical of his boss's customers: proud of her new offspring, eager for "professional" pictures of her six-month-old son. Phil had tinkled bells, made faces at the brat, snapped his camera for more than an hour.

"The proofs," he had told her, "will be ready for you to look at sometime next week. Mrs. Anderson, is there any day not good for you?"

She had said not Thursday, they'd be in Scranton then. Today was Thursday. Having been in the Anderson home photographing the baby long enough to know the layout, Phil had already decided to break through a rear porch door surrounded by shrubbery. It took him only twenty minutes. Only a fair-tomiddling score. Then he was on the road again, on his boss's business.

The Gerber apartment-house homicides, the Scarsdale caper, and the Anderson break-in—three different types of burglaries, with varied approaches and ruses. How were the victims chosen? In the Scarsdale case, Mrs. Langley's home was specifically pinpointed but only because her purse was stolen and her keys were in it. In the apartment-building homicides, the victims were obviously selected at random. Mrs. Anderson was not sought out by the photographer; if she were going to be home all week Phil Rosen would not have known when the house would be unoccupied and would have had to deliver the proofs, collecting only his meager commission.

Let's bear this principle in mind: Except for the known wealthy, those in professions or businesses dealing with precious articles, the collectors of art, rare coins, and stamps, or celebrities in sports or the entertainment field, burglars rarely single out specific people as their "marks." Thieves are far more likely to opt for favorable appropriate *conditions*. Even the most skillful burglars, and certainly the amateurs, are alert to any circumstances or situations that will enable them to get the job done quickly and easily, then to get away without being apprehended.

How you invite burglars

Among the multiple conditions that initially lure burglars is the house left for long in complete darkness. So, clearly, is the house—its owners away on vacation or on an extended trip—whose lawn is uncut or dried out for lack of attention, the air conditioners turned off in summer, the trash cans lying around empty day after day, the driveway unshoveled after a winter snow and no car tracks visible—all clues to availability.

In an apartment house without a doorman, an accumulation of letters in a see-through lobby mailbox tempts the burglar; he notes the name on the box, consults a phone directory, and calls the number to make certain no one is in. (The rapist-burglar looks for a female first name.)

If you use only a spring latch when you "lock" your apartment or house door, merely slamming it shut, a burglar can open it within seconds.

You may carelessly leave your keys in the ignition of your car when you park it in a supermarket lot or in a theater-district garage. If your house keys are on the ring a thief can have them duplicated, then return the ring to the car. Having the license number, through the Motor Vehicles Bureau the burglar learns your address. A car parked in the long-term lot at an airport tells him you'll be away from home for a while; if yours is a Cadillac or other high-priced car, he knows your possessions will probably be worth stealing.

You lure burglars, too, when you place an advertisement to sell something expensive like a "magnificently cut emerald and diamond ring," listing your phone number and stipulating that calls will be received "only after 7:30 P.M." The burglar guesses that you probably have more valuables at home. He phones first to have you give him your address. He knows your house or apartment will be unoccupied until 7:30. His path is clear.

A more palpable invitation often comes when a homeowner leaves a note, perhaps under the brass knocker of the front door, that reads: "Back at 4:30." If the thief cruising in your neighborhood sees it an hour or even a half-hour before then, he's confident that he has enough time to accomplish his mission.

You're excited about a forthcoming trip and blurt it out (including the date of your departure) to your hairdresser or favorite bartender; or you tell your service garage man to check out your car for a long trip. It's not at all unusual for one of these people to pass the tip along to a local break-in artist for an appropriate reward or even a share of the proceeds from your losses.

And there are master burglars who frequent certain places to line up their "marks." Most common are nightclubs, fur districts, and jewelry exchanges, all potential sources of pertinent information about who and where to "hit" for worthwhile loot.

Ruses, stratagems, and subterfuges

The talented professional burglar can dip into an armamentarium of tricks, techniques, and ploys, ways to spot a vulnerable lucrative prospect. Besides such tipsters or "fingers" as bartenders and beauticians, he often cultivates unscrupulous jewelers (who also may serve as fences), waiters, hotel bellhops and clerks, janitors, domestics, and ladies' room matrons in better restaurants (for names of women known to have good jewelry). He'll steal a passkey from a building superintendent. In newly constructed buildings he'll pick up keys from construction workers.

He tends to select victims the way a fisherman chooses a likely place to fish. So few people make use of effective security devices that in most cases the choice of the victim is synonymous with determining who will be away from home. To learn this, the burglar reads obituaries, interested in the time of the funeral; social columns (the Johnsons are off on a cruise); wedding announcements (to steal the bride's gifts during the church ceremony); auction announcements; the business pages (to learn of executive transfers that will mean an empty house while the family seeks new quarters in another location). All for leads to big scores.

So-called celebrity burglars buy copies of *Variety*, bible of the professional entertainment world, or subscribe to *Celebrity Service* bulletins. They learn when topflight entertainers come to town and what hotels these stars will check into. Notoriously careless, celebrities usually come loaded with jewels and often with considerable cash. One celebrated bosomy actress, on being advised after a burglary to substitute her genuine jewelry for "paste," angrily retorted: "Nothing about me is false!"

Chronic housebreaker Charlie Cheny acquired the habit of waiting outside garages in the theater district until he spotted a prosperous-looking couple. Jotting down the number on the car's license plate, he took himself next morning to the Bureau of Motor Vehicles. "A Lincoln with this license number," he reported, "sideswiped me, and I'd like to have the owner's name and address." For a small fee Charlie could then zero in on his next target.

One enterprising burglar teamed up with a salesman whose company sold wall safes to wealthy customers. The break-in did not occur until weeks after the safes had been made comfortable in the residence.

Artie Keene confidentially tells a doorman he's a private detective and hands him a ten-dollar bill. Artie says he's checking on a certain tenant's wife in an upcoming divorce case. Getting chummy with the doorman, Artie learns when the woman is usually home, when she has visitors, when she goes out, whether a maid stays in. Little by little Artie acquires the information he needs to invade the apartment at a convenient hour.

At one time—and the gimmick is still used—a couple would be enticed out of their home by free tickets to a hit show. As Freddy Miles, a long-time burglar, worked this ruse, he stole a car from the Smiths, who had left the keys in the ignition, then returned the car to their street the next day. He'd leave two hard-to-get theater tickets on the seat. With it was a note apologizing for having "borrowed" the car in a family emergency. "Please accept these tickets in compensation for any inconvenience I may have caused you." The night the Smiths enjoy the play their house is ransacked.

There are thieves who drive around a suburb on summer Sunday mornings, on the lookout for families filling their cars with picnic baskets, fishing gear, and the like, obviously off for the day away from home. An hour or so after the family leaves, the burglars tackle the empty house. (A tip for these families: Planning a picnic? Load up your car in the garage, out of sight.)

On a hot summer day thieves may drive around, park, and peer up at windows in apartment buildings. Are they looking for shades fully drawn as a sign no one is at home? No, they're hoping to find no water dripping from air conditioners. No drip and chances are the apartment (or house) is worth further casing. (A tip for you: In your absence leave air conditioners running low. Especially important in private houses.)

A stratagem similar to the one employed on Mrs. Langley of Scarsdale involves a phone call to the lady of the house, usually in a suburb. A woman's voice reports that the housewife's son has been suddenly taken ill at school. "You'd better come and pick him up." In her absence, of course, the burglary is committed.

When a targeted domicile happens to be occupied, if challenged, burglars assume a wide range of guises and have a ready alibi. They may claim to be a real-estate salesman or appraiser, a decorator, a charity solicitor, a representative of a gardening service. Wearing an appropriate "uniform"—in coveralls with a box of tools—they have posed as telephone or utility repairmen. In a newly constructed home, burglars have identified themselves as electrical inspectors.

People tend to accept strangers at face value. Some years ago, in the film *No Way to Treat a Lady*, a killer found easy entry into women's apartments by feigning sundry identities, one of which was by wearing a Roman collar. Before that movie was produced, a notorious con man had long used a priest's guise to bilk New Yorkers into contributing to his favorite charity: himself. After the film was shown, police arrested a burglar who admitted the movie had inspired his priestly uniform. There are even female burglars who dress as nuns, some of them accompanied by children as they solicit door-to-door.

Recently, in a middle-class residential section of New York, a well-dressed, twenty-seven-year-old woman posed as an Avon saleswoman arranging parties of potential purchasers of Avon products. After having ripped off at least twenty-four homes, chiefly for cash, jewelry, and items like cameras, she was finally traced through a rare, century-old violin she had stolen.

Your shrewd career burglar knows that most people will accept from a stranger any form of "identification" as long as it's printed, especially if it contains a photograph and is laminated. The stranger will then be admitted. After the robbery, the victim tells the police: "I remember the man's name, it was James Brown, I saw it on his identification card and he was with the city Building Department." You can be sure that department had never heard of an employee named James Brown.

"On the blind"

Most burglaries are committed without previous "casing" by thieves who have little or no information about their intended victims. Police refer to the practice as "working on the blind."

The least experienced burglar, often a teenager, prowls around, trying doors and windows. "Door shakers" wander through apartment buildings and hotels—usually between 1:00 and 3:00 P.M.—until they find a door that's unlocked. If a house or apartment is appealing, a burglar notes the resident's name and address, then phones the house to make sure no one is there. Does someone answer his call? He apologizes for his error in dialing, never just hangs up without replying.

Burglars like Sally Cracow, a dumpy, motherly type, would go through an apartment building on summer weekends, particularly on holidays, and merely try doorknobs. If Sally found an unlocked door she would open it and listen, while looking for anything of value just inside, a purse for example. Sometimes, if she heard someone moving about in another part of the apartment, she would quietly enter and steal whatever she could. If the person approached while she was at the door or even moving into the apartment, Sally would call out: "Yoo-hoo, Ethel?" Caught, she'd mutter the name of another tenant she was allegedly looking for and depart.

One of Sally's favorite stunts was to ride a city bus on summer days looking for fires, accidents, or anything that was likely to draw people to their windows. Finding this bonanza, she'd leave the bus and enter a building, to try doorknobs. Toward the end of her long career, Sally, arrested, told the detective who booked her:

"People. Incredible. You wouldn't believe how many don't lock their door at all—even when they go out!"

Most thieves working at random enter a high-rise building in the early afternoon and head for the upper floors; generally, apartments there offer better pickings and are less traveled. These burglars prefer to approach a door near a fire stairwell where they can slip out of sight if they are disturbed. Also attractive are apartments situated in a cul-de-sac, not readily seen from the rest of the corridor.

If the locks on a door appeal to him, the burglar listens for sounds from within. Silence leads him to ring the bell. If an occupant answers, he will—like Sally—know the name of another tenant on a lower floor, a name he has memorized from the lobby directory or mailboxes.

It's in the careful choice of easy and highly worthwhile "marks" that the best of adroit burglars, the fastidious operators, exploit often surprising, ingenious ruses. Consider the artful maneuvers of the intrepid lady larcenist known as Karen.

Karen and the poodle ploy

Midmorning and midafternoon most days, a well-dressed statuesque blonde in her early thirties walked her two poodles along Manhattan's upper-bracket Park Avenue. Karen Dosche, outgoing and loquacious, paused to chat with other dog owners out on similar strolls. She easily made the acquaintance particularly of neighbors in the building where she had recently sublet an apartment.

An accomplished con artist, before long Karen managed to get invited to the apartments of several of the dog lovers. Acutely curious and observant, Karen learned about the jewelry her new friends had acquired and she had a good eye for valuable art. Adroitly she found out at what hours, and on what occasions, each neighbor would likely be away from home.

Karen's partner in assorted larcenous games—ranging from swindles to shakedowns of men she dated—was Ted Farly, a versatile crook. They recruited an aging burglar, "Shorty" Goldin, for his valuable connection with reliable fences. None of the trio, however, was adept at picking locks, and a thief forcing doors on Park Avenue doesn't last long. They intended to have a master key made for all of the apartments in the building where Karen lived. Through Shorty, who had a wide circle of contacts, they enlisted the services of a less-than-scrupulous former locksmith, Wally Brown.

Visiting Karen's apartment one evening, Wally unscrewed the cylinder from her door's lock. By dismantling the one cylinder, he was able to cut a master key for all the apartments in the building. Through a process of elimination, in Karen's kitchen he hand-filed keys to the various depths of the pins, then in the hallways tried the key in other apartment doors. He soon found one that worked for all, put Karen's cylinder back together again, and reinstalled it in the lock in her door. Wally's role performed, he was paid off.

From that day on, Karen was vitally interested in just when her chosen neighbors would be away. At propitious times Ted and Karen phoned a victim's apartment to make certain it was unoccupied. A call to Shorty, using cute code words, fetched him in a taxi. With the master key he easily entered an apartment and leisurely burgled it, choosing the valuables Karen had heard about and anything else of value. Since he was able to bring the loot to her apartment, Shorty didn't hesitate to steal paintings and other bulky items, thus avoiding being questioned or stopped by the doorman.

Curiously, once or twice Shorty returned with empty pockets. Suspicious, Karen subsequently joined him in his forays, to keep him honest. In two months the team burgled seven apartments.

Frustrating the burglars was the foresight of some tenants—several of those Karen had most hoped to "hit"—in installing an additional lock on their doors. Any auxiliary lock would obviously not be in the house master system, so Karen's team was thwarted there.

The end came for the team just two days before Karen's sublease was up. The ever greedy Ted Farly happened to be picked up by cops for breaking into cars parked at a beach in Brooklyn, but not before a wild automobile chase down the Belt Parkway. In the trunk of his car they pulled out a set of golf clubs bearing a tag marked "Franklin Marshall, Esq." and the address of Karen's building.

The rest was easy. Loot from the last apartment the team invaded was found in Karen's bedroom. Shorty and Wally the key man were rounded up. When Karen was escorted to the slammer on that first night she asked if she could make one additional phone call. She wanted to beg a neighbor in her building to take care of her poor abandoned poodles.

PRECAUTIONS TO TAKE

Considering the various ruses and artifices apt to be employed by professional burglars, bear these cautions in mind:

► Be wary of too-inquisitive casual acquaintances. Don't let them know when you're likely not to be at home; don't mention any valuable possessions you own. Keep such information, too, from your beautician or bartender.

► In addition to the regular house lock on your front door, install an auxiliary pick-resistant lock.

► Separate your house key from you car-key ring.

► If any publicity has been given to your absence—or logically implies your absence—alert your trusted neighbors.

► Never admit a stranger until you have positively checked his credentials. Are you at all suspicious? Call his employer and verify what he claims to be—before you let him in.

► Be skeptical of "free" theater tickets or any other odd ploy that would take you away from home.

► Never invite burglars by leaving evidence that your home is unoccupied. Have the grass cut, leaves raked, snow shoveled. Keep some lights on, particularly in a bathroom, all night.

► If you're going to be away from home for an extended period, see that someone you trust picks up your mail or have the post office hold it for you. Must you tell tradespeople you'll be away? Say that a relative will be staying in your apartment or home.

► Has someone stolen your purse containing your house keys, identification, and address? Change your door locks at once.

► In a restaurant, many women tend to place their purse on the floor beside their seat—where a thief like "Jane Foster" can pick it up. Always keep your purse in your lap or on an empty chair beside you. And always bury your wallet at the bottom of your purse.

► Living in an apartment, never buzz back a door opener at your building's entrance unless you know, through the intercom, the identity of the visitor.

► If someone in your family dies, never include the address of the deceased in the obituary. It may be visited by a stranger during the funeral.

► Are you involved in a family wedding? If the gifts are at the bride's former or new home, don't include her address in newspaper announcements.

► Keep your door locked even when you are at home.

► Suppose a tradesman asks what's *not* a good day for delivery? Get him to tell you when he can come. If on that day you plan to be away, say "My sister will be here but I won't" and give him another time when you will be at home.

► Has there been some publicity about your recent purchase of an unusually valuable item? Arrange for extra security.

► And beware of strangers, such as affable lady dog walkers, who become too curious about your home or apartment.

Chapter 2 Modus operandi: How burglars break in

SHORTLY before noon one bright morning in August, Mel Hefner ambled out of his comfortable two-story home in the borough of Queens and nodded amiably to two of his neighbors tending their baby carriages. Mel was tall and slim, his long limbs agile, his angular face ready with a smile. Despite the eighty-degree heat, he wore a dress shirt and tie under his neat sport jacket, his trousers sharply pressed, shoes glistening with a fresh shine. No outward reason was apparent to regard Mr. Hefner as anyone but a solid, law-abiding citizen.

Carrying an attaché case, Mel sauntered halfway down the block to his car, a blue Mercury. He placed the case on the front seat, slammed the door, and walked down his street carefully studying every parked car at the curb. Crossing the next intersection, he continued his patrol through the next street, reassuring himself that no one was seated in any of the cars—particularly young men over five-foot-eight.

Mel Hefner was doing his thing.

What happened that summer day is related by Robert McDermott, at the time Detective First Grade attached to the NYPD's Safe, Loft, and Burglary Squad.

That day we were surveilling Hefner, hoping to catch him in the act. He was known to us as a crack burglar but we had no evidence against him for any of his break-ins. After all, few victims ever see their burglars. Thus far he had frustrated us, and we had been tailing him in a cat-and-mouse game for weeks.

That August morning I was sitting in my unmarked car a half block down the street from Mel's house. Watching him in my rearview mirror as he approached, I slid down out of sight. My partners that day, Dick Weber and Tom Sullivan, were parked around the corner.

Mel drove along the streets of his favorite hunting ground, Forest Hills. Through his mirror he watched cars behind him, pulled to the curb, sat there for a few minutes while he studied cars passing by, then drove on. Eventually he parked in the only available spot on a block lined with middle-income high-rise apartment buildings. He stood beside his car for several minutes looking up and down the street, locked his car, then casually walked—attaché case in hand—to the nearest building. Mel was working "on the blind," with no foreknowledge of his intended victims.

The building he chose had no doorman on duty during the day. If there had been one, Mel would have walked calmly past him, confident that his respectable appearance and poise would get him by any doorman. Mel would smile and hum as he pressed the elevator button for the uppermost floor. He knew that the higher up you go, the more rent tenants pay, so those floors offer a better chance for a worthwhile "pocket score" in cash and jewelry.

We knew Mel would be ringing doorbells until he found an apartment that did not respond. My partners and I waited in the street. After seven minutes (I timed it) Mel came out, walked briskly to his car, unlocked the trunk, placed his attaché case in it, slammed the trunk down, and stood there for a minute, waiting watchfully. Then he walked to the corner and completely around the block. Returning to the building he again paused outside, then reentered and remained in it for about three minutes, finally emerging again.

From his actions we guessed he had scored and must have stashed his loot somewhere in the building. His first trip out of the building was a dry run to make sure he was safe. Now, as he reached his car, we grabbed him. The attaché case was empty —his feint, in the event he'd been watched. But from his pockets we dug out eight pieces of jewelry valued, we estimated later, at $1,500 retail.

From the breast pocket of Mel's sport jacket we lifted the following burglar's tools: a picklock, a tension bar, and a jeweler's pin vise (see illustrations on pp. 16 and 17) fitted with a minute drill bit, and two fine slivers of spring steel.

These tools made entries quick and quiet, leaving little evidence of his methods. The pick Mel carried that day was a "rake." In broad terms a *picklock* is nothing more than a piece of steel—preferably spring steel small enough to fit into any keyhole—with a slight curve at its tip and strong enough to lift the lower pins of a lock and to depress the small springs holding those pins in place. A *rake* is a slight variation; instead of a slight curve it has a series of ripples or a wavy tip (see illustration).

No house lock in common use can be opened with a pick alone; a second tool is always necessary. That tool is a *tension bar* (see illustration). The two are used in conjunction and simultaneously. The L-shaped tension bar is inserted into the keyway of the lock cylinder to apply torque (twisting pressure) on the keyway; then the picklock is inserted to lift the tumblers (pins). Without the slight tension applied by the tension bar, lifting the pins will accomplish nothing. Mel had to be able to keep those tumblers up in position against the pressure of the small springs. By doing this he lined up the pins by feel, just as you line them up using your key (see cylinder illustration).

Occasionally Mel's highly touted skills failed him. Encountering a cantankerous lock he would have to resort to a technique known as *drill-and-shim*. For this he needed the small hand tool we found on him, known as a *jeweler's pin vise* (see illustration). Fitted with a 1/64-inch (or smaller) drill bit, Mel could drill the necessary hole in the cylinder in about ten or twelve seconds. After drilling through the collar of the key plug (see cylinder illustration), he removed the drill. Into the hole made he inserted one of the two slivers of spring steel (the spring from a watch is ideal). Then he lifted the pins with his pick and slipped the spring steel farther and farther into the hole until he had all the pins lifted, the shim holding them in place instead of the tension bar.

Besides his assorted tools Mel had in his pockets four business cards in different names. They identified him variously as an insurance salesman, an interior decorator, a telephone company business agent, and a collection agency representative.

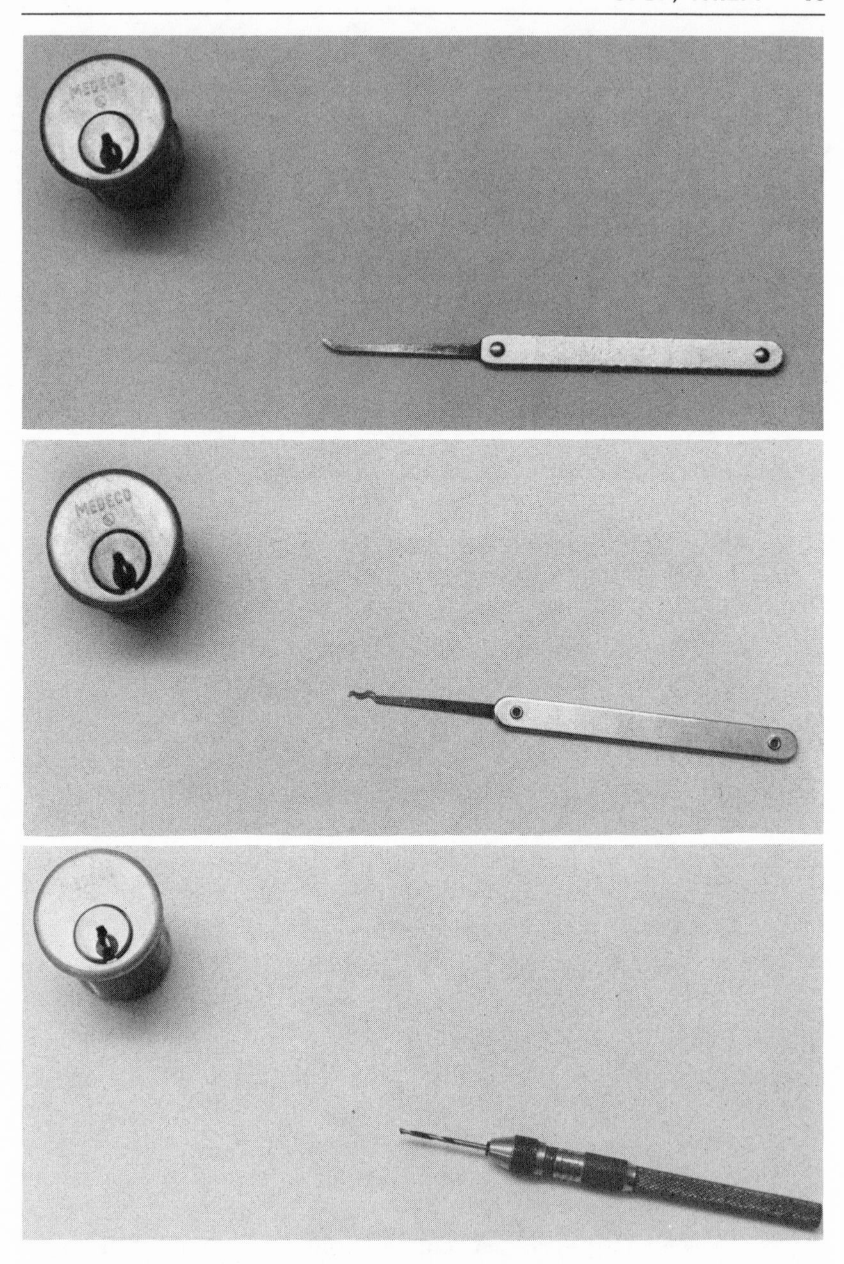

Top A picklock enables a burglar to make quick and quiet entry, shown here in contrast to typical lock cylinder, indicating tool's size.

Center A "rake" is a slight variation of the picklock.

Bottom Jeweler's pin vise, used for cantankerous locks.

Predictably, Mel protested his innocence. Ordinarily, when we got a pick man we'd haul him to the nearest precinct station and have to wait until some distraught victim reported a burglary because there was no outward sign of a break-in. Since Mel had the pin vise and other burglar's tools in his possession, this case proved easier.

I went to the upper floors of the building. There I checked apartment doors for a minute hole drilled just above the keyway (see illustration on p. 19). On one door fitted with three locks I discovered one of the cylinders had been drilled. Apparently this expert burglar had successfully defeated two of the lock cylinders but encountered trouble with the third and had to drill-and-shim that cylinder. The process probably took only two minutes of the seven Mel had spent in the building.

Like all of us, thieves are creatures of habit. Dressing in the morning you put either your right shoe on first or your left. Invariably. Burglars are just as repetitive. Their habitual methods of operation—how they break in, the tools they employ, what they do when inside premises, their predilections—all add up to what police refer to as a burglar's *modus operandi.*

Tension bar, used in conjunction with a picklock.

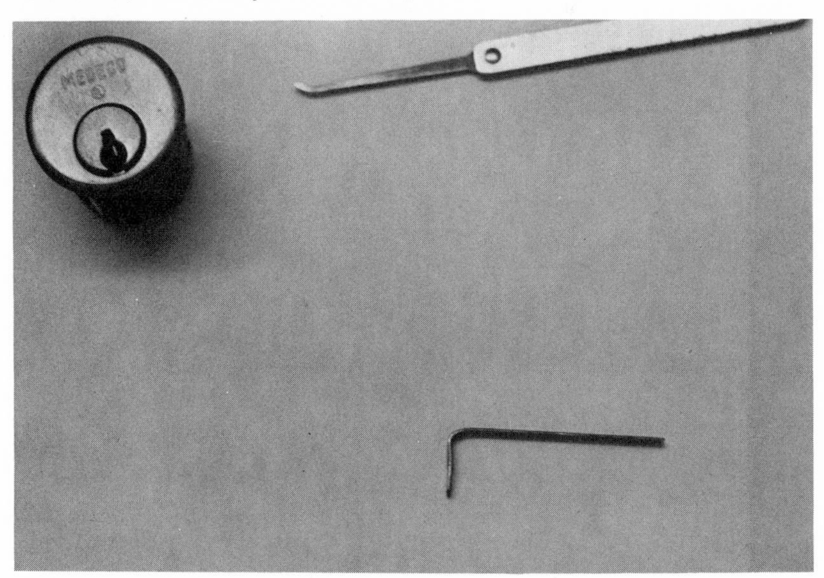

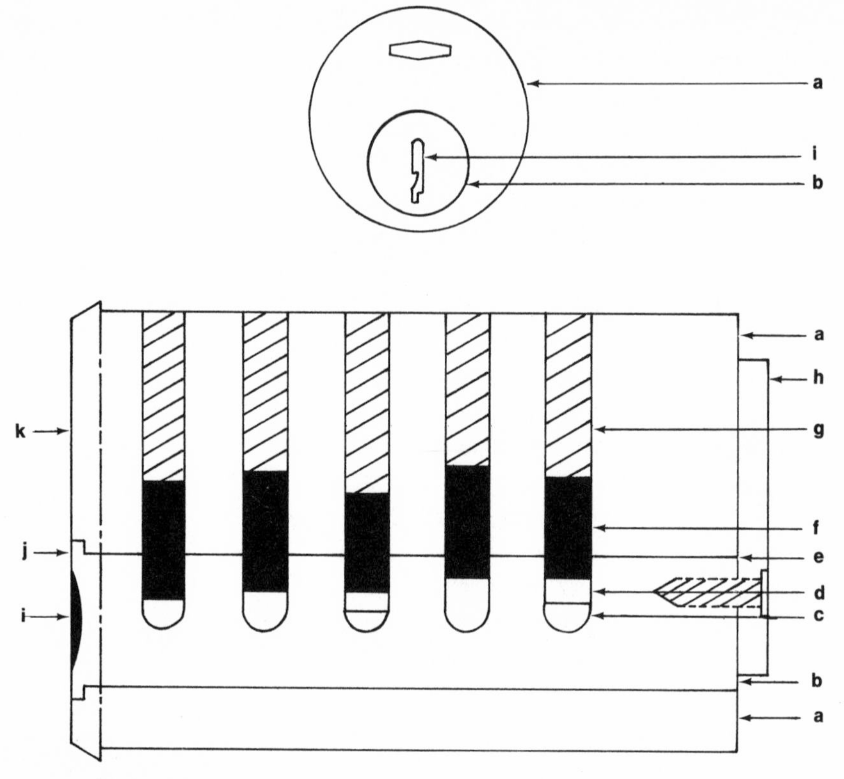

A pin-cylinder "skeleton" showing its various parts: **a.** cylinder shell; **b.** barrel (plug); **c.** follower (lower pin); **d.** master pin, or wafer, allows two different keys to actuate; **e.** shear line; **f.** driver (upper, or locking, pin); **g.** spring; **h.** cam, which engages to withdraw bolt on actuating cylinder; **i.** keyway; **j.** point for drilling in "drill-and-shim" technique of defeating lock; **k.** point for drilling all the way through the cylinder.

In this age of specialization burglars can be as precise in their specialties as doctors are. Certain types of thieves focus exclusively on jewelry, others on furs, bankbooks, or art. There are hotel rip-off experts, private-home bandits, and the snobbish species who will go only for penthouses. Burglars may insist on working solo or with a partner or partners.

The M.O. will depend not only on conditions at the target but also on the burglar's particular skills, personality, connections—or pride. Some years ago a Brooklyn, N.Y., pickman was being utilized by the Mafia for some highly specialized assignments that he felt were commensurate with his talents. One night his boss got drunk and ordered the lock artist to pick open the door of a tavern so that he could have another drink. The pick man, acutely offended that he would be exploited in this menial way, shot the Mafioso to death.

Some professionals prefer office buildings, others work only in stores, lofts, warehouses, or church rectories. Apartment thieves may do their jobs only in specific neighborhoods. And there's the versatile, usually less skilled kind who may hit any place that's convenient and vulnerable.

As for the burglar's means of entry, styles are multifarious. At private homes, back doors are a favorite, and windows that allow the intruder to be shielded from the street and surrounding homes while he operates. At apartment buildings he may enter through the delivery service door or even the front door, waiting until the doorman is preoccupied with other visitors, delivered packages, or calling a taxi for a tenant.

Other burglars will use the fire escape or go to the roof and use a rope ladder to reach the terrace of a dark apartment. The

Minute hole drilled above keyway of Sargent cylinder.

specialist known as a *cat burglar* or *ledgeman* enters through a seemingly inaccessible window by edging along ledges or stepping over from an abutment or fire escape.

New York's Plaza Hotel was once under siege by a cat burglar who operated for weeks high above the street, taking advantage of the hotel's architecture. The security men began to think of him as a Phantom of the Opera type because he was constantly being seen by guests as he crawled along a narrow ledge. Although he was chased several times through corridors and was once shot at while out on a ledge, the "cat" persisted in scoring. Early one morning he was trapped by two radio-car patrolmen who had responded to a call from a guest reporting a peeping tom outside her window. Hauled in through a window and searched, no tools were found on him. Why would he need a jimmy when no one would feel the need to lock a window opening out on space?

Distinction should be made, as we've noted, between the "crude" burglar and the proficient professional. The crude type generally makes a forcible and obvious break-in. Poorly dressed, he seldom if ever tackles better-class residences where he'd be too conspicuous. Like a scavenger he steals anything he sees of value he can carry, turning a place upside down, grabbing everything from radios to clothing and piggy banks.

He roots through your home gathering all valuables to a central spot, usually right inside the door he is going to use for his exit. There, he (and others who may be with him) pile loot from all the rooms until he feels the pressure to get out. Then he chooses the items for which he can get the most cash and will carry out as much as possible. When you find some of your property still stacked when you arrive home, it isn't there because the burglary was interrupted and you "nearly walked in on him."

Without skills the crude burglar is apt to rely on a jimmy. Often, if dissatisfied with the loot, he'll indulge in vandalism. He rarely invades a place with a person in it and if someone comes home while he's working he'll try to flee—unless the victim offers resistance. Then the crude crook will fight his way out, using any weapon at hand, including his jimmy.

Among the favorite targets of drug-addict crudes are drugstores and doctors' offices, for obvious reasons. As we've pointed out, burglars rarely select individuals as prey except for those engaged in enterprises that deal in something of particular value. To the addict nothing holds as much value as his "high." Most often the break into a doctor's office or a drugstore will be not only forcible but amateurish; these thieves are usually soon caught and put away.

Very few of the crudes have a top-notch fence, so they net little from their forays. It means they have to burgle more often, even every day of the week, especially when they must support an expensive drug habit. Making obvious, noisy breaks, they are caught more frequently.

In sharp contrast, the elite or career burglar knows a lot about locks, alarms, and other obstructions to entry, and may specialize in safes and/or vaults. Police rate him professional on the basis of his trained talents, his targets, his appearance, his careful planning for any eventuality, and the quality of his fence (receiver of stolen items). Certainly he does not "look like a burglar." Almost never is he on narcotics, though usually he's an avid gambler.

Since the pro is fully aware that if caught he'll receive the same sentence for a large haul as for a small one, he invariably goes after a big "mark." With a sizable hit, he need not work too often, so he has less chance of getting caught.

Rather than use force, the pro is more likely to contrive a *surreptitious entry*—a term referring to a quiet and quick entrance without leaving any visible mark to indicate how he managed to do it. He rarely resorts to a phony uniform, almost never carries a gun. He is glib, brazen, inventive—and hard to catch. The best of his kind operate for years before they are finally nabbed.

Once inside, the pro works fast. First he heads for the master bedroom—a woman's dresser and a man's chest, the closets. In a man's den he goes for a safe, cash box, stamp or coin collection. The pro knows where people are likely to stash valuables. He has dug them out of record albums, socks, vases, violin cases, clothes hampers, even a douche bag. The crude burglar may not be as smart or as thorough, but he does spend a longer time in his search and will certainly empty your coffee, sugar, and flour canisters onto the floor.

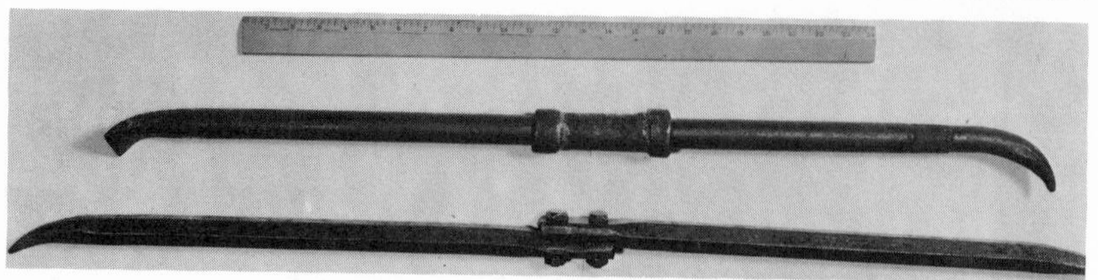
Sectional jimmy, assembled for two lengths.

Burglaries committed in residences and business places may be classed under two broad categories: forcible and surreptitious.

Forcible entry—through doors

When police report that there were "no signs of forced entry" in a burglary it means that there were no "force" marks on doors or windows and that tools such as a jimmy were not employed. A favorite of burglars, especially young ones and drug addicts, the jimmy is a leverage device like a small crowbar, packing-crate tool, screwdriver, or tire iron (see illustration). It's used most often in private homes or business establishments but not so much in apartments, partly because it's bulky and its use results in considerable noise.

Used as a lever to gain purchase for prying, the jimmy is forced between a doorframe and the door's edge, usually in the immediate area of the lock bolt. The idea is to force the bolt out of the strike plate (the metal-lined recess in the frame) or to force the bolt back into the lock. Pressure causes the wood of the doorframe to split, freeing the bolt or the screws mounting the lock to the door to be pulled out.

When a doorframe is old and dried out, the wood rotted or lightly constructed, just one jerk with a jimmy yanks the bolt out of the frame. On such occasions, even if the lock bolt is still extended, the burglar can walk in. If a lock is made of die-cast materials and not solid brass or bronze, the lock itself will fracture.

Without a jimmy a burglar may use vise-grip pliers or a small pipe wrench to unscrew a lock cylinder. The attack is relatively

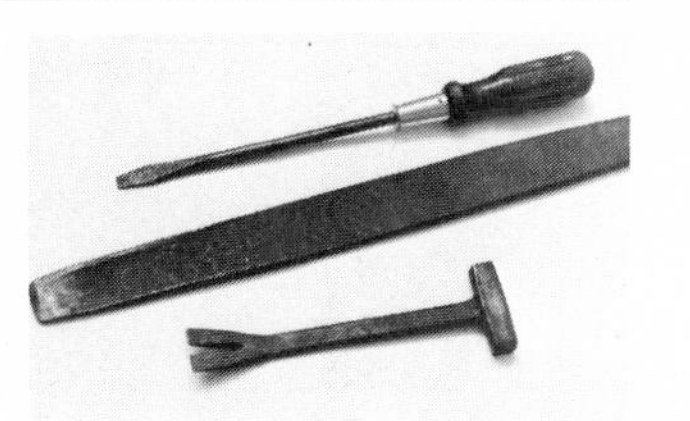

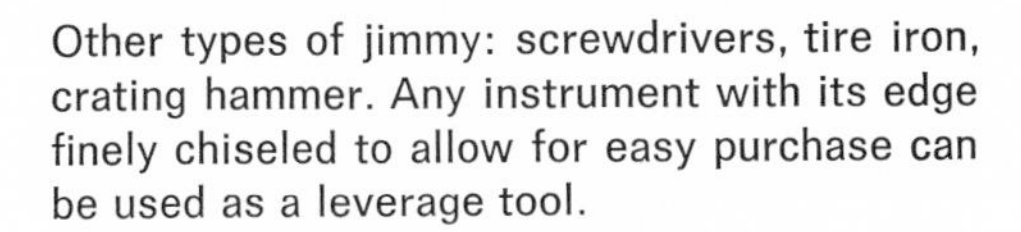
Other types of jimmy: screwdrivers, tire iron, crating hammer. Any instrument with its edge finely chiseled to allow for easy purchase can be used as a leverage tool.

silent, requires only the use of a tool easily concealed, and the break is not as obvious to passersby as jimmying. The more careful exponent of this "cylinder wrenching" will, on entering the premises, cover the gaping cylinder hole by pushing the original cylinder back into place. If the torn-out cylinder is too damaged, he will use a "dummy" cylinder. Available at hardware stores, these are normally applied to cover up a hole when you remove a lock from a door.

Key-in-knob locks (or "tubulars"), the favorite hardware of apartment and private-home builders—because they are easy to install and economical—are extremely vulnerable to forcible entry, yielding to several methods. The bolt in these locks extends only about half an inch into its strike plate (the recess in the frame) so that jimmying is easily and quickly achieved. These key-in-knobs are so poorly designed that many burglars use just a screwdriver to attack them. They press the tool against the edge of the door, aiming directly at the bolt (always set even with the center of the knob) and drive the screwdriver in with the heel of their hand. Often this action alone will force the bolt back into the lock and out of the strike.

With very few exceptions these locks in homes, apartments, motels, and most business places have an even more glaring weakness than the fact that they're easily jimmied—and it's known to any savvy thief. Again, all he needs is vise-grip pliers or a pipe wrench with which to grip the outer knob. As you know from the lock on your door, when the mechanism is "locked" the outer knob becomes rigid—you can't turn it. When you can turn the knob (unlocked) the bolt moves in and out. By forcing the locked knob to turn, it follows that the bolt will be withdrawn and that's exactly what the burglar does. Using his pliers or wrench he forces the rigid knob and presto! the bolt moves out of its locked position.

Then there's frame-spreading (see illustration). All a housebreaker needs is a portable jack to spread the doorframe, and the bolt in the lock is left hanging loose. Or he may slip a hacksaw blade between the door and the frame, cutting right through the bolt.

Rivaling the jimmy in popularity with the crude thief is the "foot"—his own foot aimed right at the lock. It is much better than applying a shoulder. He is the kick-it-in man.

Occasionally a thief will lay a punch on the face of the lock cylinder and hit it with a hammer to knock the surface-mounted "rim" (or auxiliary) lock right off the door.

When a thief is interested in a specific lucrative score the lock may be very difficult to open with a jimmy and the cylinder may be protected by a cylinder guard plate or a hardened-steel cylinder guard ring (collar) to prevent it from being pulled or wrenched with a vise grip or pipe wrench. In such a situation the burglar may resort to drilling out the locking pins of the lock cylinder or he may fall back on the drill-and-shim technique mentioned (Mel Hefner's technique). In these attacks the in-

Spreading of the door and frame.

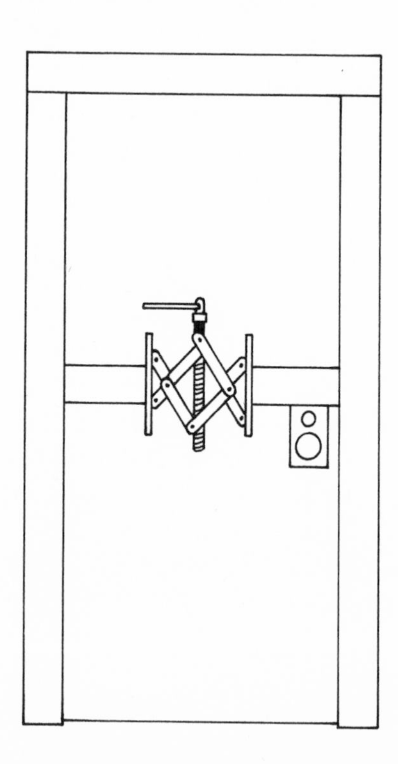

tended victim's lock cylinder will not be a high-security type. A high-security cylinder is not only pick-resistant but also has steel inserts to protect against drilling.

With drill-and-shim, the thief maneuvers the pick and shim to solve the puzzle of the lock's obstructions, dealing with each of the (generally) five pins one at a time. A skilled burglar can do the job in a few seconds. When he leaves, he removes the shim, often hiding the tell-tale hole with a silver- or gold-colored crayon.

One of the busiest exponents of the jimmy was Bernard L. Foley, a brash, handsome young Texan who used to select his victims through an escort service and car-rental agency he ran in Dallas. When Dallas got too hot for him he headed for the greener pastures of the Big Apple. There he tackled suburban homes, then brownstones and apartment houses in Manhattan. Since locks were a total mystery to him, Foley relied on his dependable jimmy and vise-grip pliers to tear cylinders out of back-door locks or apartment doors. He would hit so hard that doors were often left half off their hinges.

At apartment buildings Foley followed a set pattern, committing a burglary about every eight or nine days. With one eye on the doorman, he kept watch at the head of the ramp leading to the building's basement garage for tenants. Many such garage doors are electronically operated, requiring the use of a card inserted in a card reader to open them. As a tenant's car arrived and the driver inserted his card, the door opened and Foley followed the car off the right rear side (out of sight), walking into the garage. From the garage he could rove at will through the building, free to choose an unoccupied apartment.

Foley's trusty jimmy and other burglars' tension bars, wrenches, celluloid, and picks represent the most common techniques of entry through doors. A variety of other methods, of course, are resorted to by criminals from time to time, some of them seemingly intricate but all simple when understood. It's not uncommon, for instance, for a burglar to break out a glass pane in a door, reach through, and unlock it by turning the thumbpiece inside. (The locks and devices described in Chapter 11 will prove effective against all such methods of forcible entry through doors.)

Clasp-type lock for window.

Forcible entry—through windows

Particularly in residences, though also in certain commercial establishments, burglars tend to break their way in via windows rather than doors. The majority of these windows are of the double-hung type and constructed of wood. Universally fitted with simple clasp-type locks (see illustration), they offer very little resistance to prying with a jimmy. The clasp lock may also be opened by a thief using a bread knife ("shove knife") slipped between the upper and lower sash. If his knife is obstructed by a metal insulation strip he can drill a small hole in the outer sash and move the clasp open with a stiff piece of wire.

Side play in the window, which permits it to be moved from side to side, is another weakness. A burglar moving the window from side to side with a slight lifting and rotary motion can work the clasp out of its locked position.

Less common than the breaking of a small glass pane in a door to open a lock from inside is the breaking of a large windowpane to gain entry. That's apt to make too much noise. Rarely is a glass cutter resorted to. In fact, as any glazier will tell you, if the burglar scores the glass with a cutter he must then strike the glass from the opposite side to have the glass break away cleanly. Small, multiple panes in a window do, however, constitute a problem; they are easily and more quietly broken by a burglar who will then reach through and unclasp the window lock.

As for casement windows, the burglar may use a pointed knife to remove the putty holding the glass in place. He selects a pane near the window catch. Needle-nosed pliers remove the small wire clips retaining the glass. With clips and putty out of the way, the pane drops into the burglar's hand. Sometimes he'll use a rubber suction cup to pull the pane out of the window frame.

In urban areas criminals who use fire escapes will attempt to enter through windows protected by the folding accordion type of barred window protection known as a "ferry gate." With some of these the burglar finds he can force the lower edge of the gate out of its track and crawl under them. The best gates have a *deep* track running entirely along the lower sill and top of the window. Some thieves who encounter these better gates too often will employ a bolt cutter to cut the shackle of the padlock used to secure the gate.

The cat burglar works apartment houses more often than private homes. His usual technique is known as the "step-over." From a fire escape or other abutment he approaches a window not normally considered accessible and he steps to the sill; from there he enters a window probably not as well protected because it seemed wholly unapproachable.

Burglars who effect entry through windows will not hesitate to steal bulky items. They usually leave through the door.

A favorite target of the house burglar is the sliding glass patio door. With few exceptions these doors are fitted with locks (and cylinders) easy to force. In many it's possible for the thief to lift the sliding door right out of its track. There's little value in blocking the track with a length of wood to prevent its being slid open. We suggest that with the door slid fully open you glue a strip of metal (or wood) along the entire length of the recess in the upper track. Use a gauge of metal that will permit the door to slide freely yet take up the slack so that the door can't be lifted up and out of the lower track. Then you can prevent the door from sliding open by cutting a length of wood or pipe and laying it snugly in the lower track with the door fully closed.

Taking such action will add immeasurably to the security of your sliding doors. If the lock on your door is obviously weak or malfunctioning (as many of them are), replace the lock through your locksmith; ask him for the best he has in "narrow stile hook bolt" locks.

Surreptitious entry

One of the most prevalent—and simplest—means of foiling a locked door is through the use of a celluloid strip, the *'loid* (see illustration). Here the thief takes advantage of the inherent weakness of the very common spring-latch mechanism of a door lock. If you have such a lock on your door and you merely slam the door when you leave without turning your key in it, you are a sitting duck for thieves. Until recently burglaries contrived by 'loids accounted for a majority of surreptitious entries in most large cities.

If a 'loid practitioner encounters a "double-locked" door (where a key was used upon leaving), he can't use his gimmick and will move on to another, less careful potential victim. Frustrated by the double-locked door he will rarely turn to forcible entry.

A plain strip of celluloid is seldom used. The burglar prefers some commonly carried article, typically a wallet-sized bank calendar of pure celluloid or a credit card. If he is arrested and charged with possession of a burglar's tool he is confident that the judge and jury will be carrying one of these cards in their wallet, so that his conviction will be more difficult to obtain. In another strategy, the burglar carries five plastic-coated washable playing cards, a full house in poker. And it will be up to the prosecutor to prove that these are not a souvenir of a Saturday night game. In hotels and motels burglars often manipulate Do Not Disturb signs for their 'loiding.

In operation the celluloid strip is introduced between the door and the doorframe, a few inches above the place where the burglar knows the lock to be. He presses on the door with his hand, high over the lock, taking advantage of the slight give in the door. Having inserted the strip without disturbing the door molding, he then works the 'loid down until he encounters an obstruction—the beveled latch or lock bolt. He knows by the feel of the bolt whether he is dealing with a beveled spring latch (see illustration) or a dead bolt. Many burglars can tell if the bolt is "alive" (spring latch) or "dead" (dead bolt) just from the position of the bolt in relation to the cylinder. If the bolt is alive he then proceeds to work the ramped surface of the latch out of the strike plate—simply by pushing on the 'loid. If the bolt is dead he knows he has something quite different to contend with and will seek his loot elsewhere.

When there's more than one lock on a door, can a burglar tell if all are actually locked? This largely depends on how far apart the locks have been installed on the door. If an auxiliary lock is mounted ten or twelve inches above the regular house lock, by exerting pressure with his hand high on the door the burglar may be able to determine whether you've locked the additional lock.

PICKLOCKS The burglar who first tries a 'loid and fails will seek a different victim unless he is capable of manipulating picklocks, today the most prevalent of surreptitious techniques. Picking a lock requires far more skill than any other type of entry. Many thieves know how to pick locks, they know the theory, they may have seen it done, but relatively few become

The 'loid is inserted between the door and molding to draw the spring latch out of its strike, thus unlocking the door.

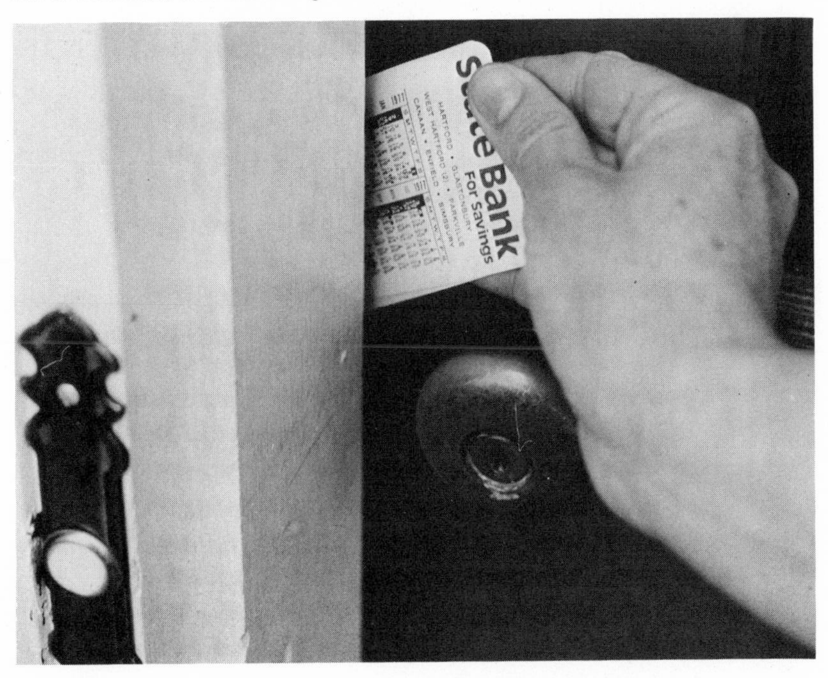

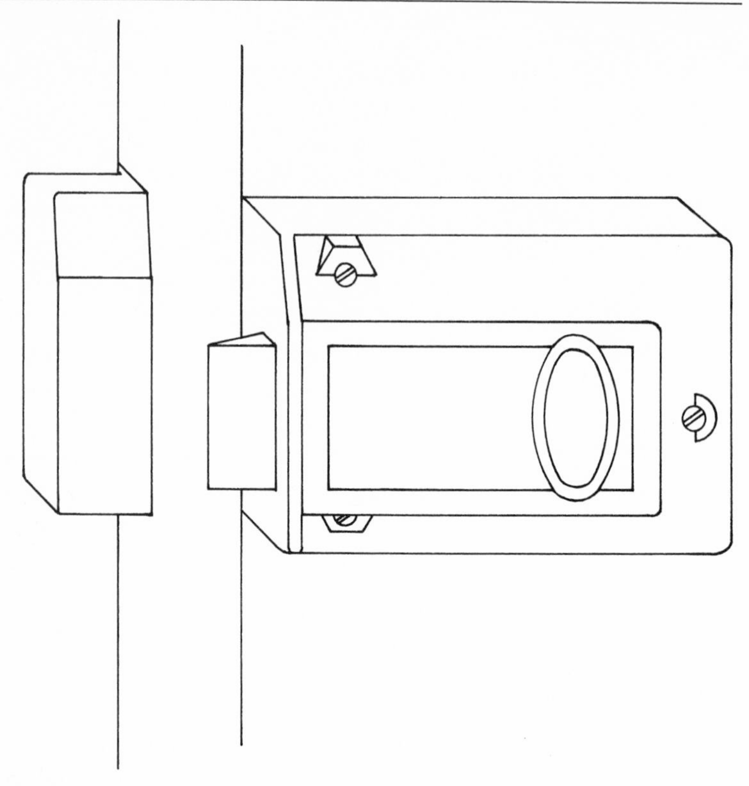

A beveled spring latch.

really proficient. But a burglar doesn't have to be the best to beat the ordinary locks used today.

Pick men are the elite of the profession. A career burglar like Harry K. practices every day, using the latest "burglar-proof" locks, which he buys from distributors and manufacturers. His fingers are uncommonly nimble and sensitive. Though Harry appears to be smooth and talks calmly, he's always nervous, working under tremendous tension, and any unexpected noise disturbs him. On the job, if a pick-resistant cylinder takes him too long to defeat he'll walk off.

The really dedicated pick or 'loid man wears jacket, collar, and tie, so that he will blend into the environment he invades and not call attention to himself. To be forewarned of an apartment tenant unexpectedly walking in on him, the burglar—after he has entered—will often plug the keyway with a match, toothpick, or small piece of metal. If the tenant arrives while the thief is at work, he'll be unable to insert his key in the lock and inevitably will leave the corridor to seek aid from the superintendent or a neighbor. That gives the criminal a chance to vanish.

Pick men rarely steal anything they can't stuff into their pockets; hence, the "pocket score." Those who are inclined to steal furs enter with an empty new suit box equipped with handle and fresh string.

As he makes his exit from a purloined apartment the smart crook will "talk" to someone presumably inside; that's to allay possible suspicion by anyone who may be in the hallway, perhaps awaiting an elevator. If there's no one outside, the pick man removes the plugs in the keyhole. The objective is to delude victim and police into assuming the burglary was an inside job.

Searching an apartment, a veteran pick man, Johnny Fiore, would look for a bank passbook with a considerable balance. He'd steal the passbook, ignore a small amount of cash lying around, and leave the place otherwise undisturbed. If the victim missed his or her passbook, he (or she) was unlikely to believe it was stolen. Johnny would also pick up a few household utility bills and a sample of the victim's signature or handwriting. At the bank Johnny would have a girl friend submit a forged withdrawal slip with the passbook—and the utility bills if identification was called for. The victim's account was never entirely depleted, so as to forestall suspicion by the teller.

MANIPULATION KEYS The overwhelming majority of locks are operated by a mechanism known as a *pin cylinder*. This is the device you insert your key into and rotate to actuate the lock and remove a bolt. Virtually all modern locks make use of a separate removable and changeable part that is screwed into or attached to them by mounting bolts. This part is the "cylinder," beloved of pick men and thieves who use "manipulation keys."

Popular myths to the contrary, there are no master keys that will open just about any lock. There are, however, manipulation keys. If you live in an apartment building or in a private home built as part of a development, the odds are that the cylinder on your door is "mastered." This simply means that your door can be opened by at least two different keys—your own and the building superintendent's (for emergencies), or builder's, master key.

Large office buildings and hotels have several keys—all appearing different from each other—that will open any door, usually for cleaning employees. Each of these keys, when used in a cylinder, lifts the pins to a different level, yet all will operate the lock. Thus, if a thief wants to pick this mastered cylinder he can succeed by lifting the pins to a variety of levels. The job is easier for him than if he is going after a cylinder with pins that will only unlock at one specific level (as in the case of an unmastered cylinder). For the same reason, a burglar can take any key that will fit into the cylinder keyway and file the sides of it down so that it moves sloppily in the cylinder. Then he manipulates it, moving all of the pins up and down, hoping to hit one of the right combinations to unlock the door. It's a very iffy situation, possible only on mastered cylinders.

Manipulation keys do not constitute much of a threat to your security—but master keys and mastered cylinders do.

Master keys may be "borrowed" by a criminal from a dishonest building employee and copied. They may "disappear" from their usual hook in the vestibule of the apartment building where they are hung for use by the doorman when a tenant is locked out. They may be stolen from new residential construction sites. In short, you never know how good the control is over such keys.

Some of the highly pick-resistant cylinders now available can be mastered if necessary but these will still not be picked with the ease of the standard types.

There are tales of master thieves who can "read" a key—who, with a mere glimpse of a key, can determine the cuts in the key and later duplicate them from memory. It's been done at locksmith conventions as an entertainment, usually with simple warded keys (the small, flat type used in clubhouse locker rooms). However, no burglary we know of has ever been done this way.

The chain bolts

This device, attached to the inside of a door, is designed primarily to safeguard the person within an apartment or house rather than to protect property. Chain bolts can be forced easily, or they can be removed from the outside by a skilled burglar without evident force, by one of several methods.

Let's say you have the typical chain lock found in most apartments and private homes. After 'loiding or picking the door lock, a burglar can slip a hand and forearm inside and press a thumbtack into the wooden door just beyond the chain. He then loops a rubber band from the tack to the end of the chain where it rests in the slide. Withdrawing his arm, the thief pulls the door closed. This relaxes the chain. The rubber band now pulls the slack chain to the channel's end. There, still being pulled by the rubber band, the chain drops free of the slide and the intruder steps in.

If the door is metal-clad the burglar uses a sticky tape instead of a thumbtack. When the chain bolt is so installed that it prevents his getting an arm through the opening, the thief reaches through with his fingers and attaches a hook (a paper clip is ideal) to the last link of the chain that he can reach, the one farthest from him and nearest the slide. Attached to the hook is a long piece of string. Holding the door open as far as he can, with the string now in his hand outside the door, he reaches up and draws the string over the top of the door toward the hinge. He closes the door, taking the pressure off the chain, pulls on the string, and draws the chain out of the slide.

Chain bolts that allow you to lock them when you are leaving your home can usually be opened by either of these two means. In addition, the locks used on them are the same as those found on desk drawers and can be picked very easily.

After entry is made

Once the burglar has taken advantage of your lack of security his actions will follow a pattern consistent with his method of entry.

Pick men, key men, and those using celluloid strips, having effected an entry without an obvious break, will plug the cylinder and invariably head for the master bedroom. Here, initially at least, they cause little upset. They know what they are looking for and where the "pocket score" is most likely to be. They check the surfaces of furniture—dresser, dressing table, night table, vanity—probing for a jewelry box. If they fail to find one in open view they quickly go through drawers. Thieves know that what they are seeking will be in some sort of container, even if only a chamois bag, and they feel through the clothing for this solid object.

If the loot is not located in drawers, the burglar will then rummage on closet shelves or drawers built into your closet. In most cases the score will be made from one of these obvious hiding places. If he finds only a single piece of valuable jewelry during his search, he may then start to toss drawers and contents on a bed. He knows that if he has found one piece there should be more; you must have it scattered about, so he ransacks.

Failing to locate jewelry but coming upon a safe, he will probably depart, disturbing as little as possible, to return at a future date equipped and prepared to either attack the safe or force you at gunpoint to open it.

The criminal who achieves surreptitious entry has, of course, caused no apparent break-in. The victim will probably feel certain that the thief was "someone with a key" or someone "who knows me"—because that thief knew "exactly where to find my jewelry."

PRECAUTIONS TO TAKE

To help frustrate the most common of the burglar's M.O.'s, here's what you can do:

► Leaving your apartment or house, keep a radio playing at normal volume; in summer, don't turn off your air conditioners. On-the-blind thieves like Mel Hefner, listening outside your door, are apt to believe someone is at home and will pass you by.

► Don't rely on a lock that you just slam whenever you leave. Use one that requires a key to lock it.

► In a private home, office, or business establishment where an alarm device would be appropriate, install a dependable system. (For choices, see Chapter 11.)

► Never rely on a key-in-knob lock for security. To keep forcible-entry burglars out of your apartment or house, your first step is to have a locksmith install a jimmy-resistant vertical-

This cylindrical deadlock has a one-inch hard-steel bolt and security collar. Used as an auxiliary for residential or commercial doors.

bolt deadlock or a good "cylindrical" dead bolt (see illustration) on your exterior door. A dead bolt is a heavy, unbeveled bolt that must be turned into its strike plate with a key or knob. It can't be forced open with a 'loid or a knife. Since it is longer than the spring latch, it resists a jimmy. And because its keyhole is not in the doorknob, it can't be broken open with a pipe wrench.

► When you have keys made in one of those "while-you-wait" shops, watch to see the keymaker does not make one for himself.

► Chain bolts on doors are best used in conjunction with a peephole to identify visitors unknown to you. Make certain the chain is short enough so that an arm can't be inserted to grab at you. Screws fastening the chain bolt should be strong enough to resist a hefty shoulder striking the door.

► If you have a glass door be sure to buy either a vertical bolt deadlock or a good cylindrical dead bolt with a double cylinder in it that requires a key to be used from the inside. (For more information, see Chapter 11.) If you elect to use a surface-mounted deadlock, this must be installed with one-way screws (designed to permit only tightening) so that the lock can't be removed by the burglar reaching through your broken glass with a screwdriver. Then, if the glass pane is broken out, the thief won't be able to open the door.

► To prevent break-ins through windows, use key-controlled window locks. Or "pin" your windows by drilling a hole through the inner sash and partly into the upper sash (see illustration on p. 37). Insert a tenpenny nail or pin in the hole. The hole should be made at a slight angle upward (the back of the drill lower than the front)—about five degrees; this will prevent the pin from being dragged out of the hole if a heavy jimmy is used. Drill out additional holes for key-controlled locks and the pins so that your windows can be partly opened for ventilation. If you decide on keyed locks for your windows, order them from your locksmith keyed alike—and hang a key near every window in the event of fire but out of reach of someone outside.

► If your main-floor doors and windows are well protected a burglar may try to enter through a basement window. Most often such windows are hinged at the top. You can easily secure them by installing eyebolts in the window and frame, then slipping a padlock through them. Or use a single-hinged hasp and padlock.

► Do you live in a house? Clear away shrubs that prevent passersby and neighbors from seeing your windows. If burglaries have been committed in your area recently, install lights at the corners of your house to illuminate vulnerable windows outside at night.

► Give serious thought to where to hide your valuables. Easily detected by professional burglars, for example, are jewelry buried in the bathroom clothes hamper, or beneath shirts in dresser drawers. Expensive jewelry and other valuables should be kept in a safe-deposit box.

The dangerous burglar

In pinning down precisely what makes a burglar the criminal he is, one characteristic remains unchallenged: the "compulsive quality" cited by behaviorists. That element is found in just one other branch of lawbreaking: sex crimes.

Clinton T. Duffy, the celebrated ex-warden of San Quentin prison, held that the burglar's primary motivation is sexual, that the stealing—the score or gain—becomes secondary. In other words, if the burglar leaves a place without filching anything he may still be satisfied. Here's how Duffy's neatly packaged hypothesis goes:

Entering a house or apartment, Burglar Joe often finds slim pickings and he gathers up whatever is handy. But there's more to catch his fancy. In Joe's daydreams the place he breaks into at night is deserted except for a female sleeping alone. And in his fantasy he sneaks into her room and stands by her bed. She wakes, sees him, smiles languidly, and opens her arms to him.

His fantasy becomes reality—until the woman opens her eyes.

Instead of smiling, she screams and tries to evade Joe. If he is determined to get what he came for (sex), he resorts to rape.

A burglar-rapist, Duffy believed, is always capable of inflicting serious physical harm. He is apt to carry a knife, a screwdriver, anything that may serve as a weapon in attacking his victims, some of whom are left badly battered.

In other words, a housebreaking rapist presumably hides under the guise of burglary. And slipping or "coming" into a dwelling ostensibly to commit burglary is erotically stimulating, a kind of aphrodisiac.

While there is an element of truth to this theory, we feel it has a flaw. In our experience, the burglar does expect his material reward to be commensurate with his effort. Too many burglars have become virtually suicidal if, after getting in, they miss the swag.

Behaviorists often refer to the professional burglar as abhorring violence. It's an extension of the "gentleman burglar" mythology. Let's face it: All burglars are knaves, their intent entirely evil—and they are all potentially dangerous.

In some burglars, undoubtedly the surreptitious entry does arouse a sexual drive demanding release. This was certainly evident in the sex slaying of career women Janice Wylie and Emily Hoffert, which we'll go into later (see Chapter 7). When

A window is "pinned" by drilling a hole through the inner sash and partly into the upper sash.

these men get sexually stimulated and a woman is available on a job, almost invariably they cruelly degrade their victim.

This sadistic compulsion was clearly illustrated by another brutal case, a little-known mystery where Detective McDermott performed a vital role in revealing an overlooked modus operandi. To protect the victim we'll call her Susan, her parents Doctor and Mrs. Clarke. This is how McDermott recalls it:

My first knowledge of the crime came through a telephone call to my office in the Safe and Loft Squad early one Monday morning. Frank Gottlieb, a highly respected precinct detective, was sending over for my examination two lock cylinders removed from the door of a Brooklyn apartment where a girl had been raped the night before. He indicated there was no sign of a forced entry into the apartment and he would appreciate my judgment as to whether pick tools had been used.

"Do you know of any pick man who's a rapist?" Frank asked me.

"Possibly," I said. "If the circumstances are right. But it's unlikely. Usually pick men have other things on their minds. What was stolen?"

"Nothing, as far as we can tell. Mac, this was a vicious job on a fifteen-year-old kid."

"Fifteen! I know some weird bastards picking locks, but a sadistic rapist? No, I can't think of anyone. Did you say this thing happened at night? What time?"

"We can't be sure. We can't talk to the girl, she's out of it. It had to be after eight-thirty and before eleven-twenty. That's when the kid's father, a doctor, found her. The blood and semen were pretty much congealed by then."

"Blood? How bad is she, Frank?"

"Well, we think it may come to murder."

"I see. Listen, if I'm wrong, I'll apologize, but I don't see a pick man in this. Not unless the doctor is really loaded, a particular target, and the pick knew no one was going to be in the place. Did the girl walk in on him?"

"No, it doesn't look like that. She was probably doing her homework and he came in on her. Or she let him in. That's why having you look at the locks is so important. Right now we think she may have known the guy."

"Doesn't sound like any of my customers. Any of them looking for a good score would first make a phone call, to make sure nobody was there."

"The apartment doesn't look to me like there could be any exceptional score."

I shook my head, more and more puzzled.

"The doctor not a target," I mused. "A rapist who took no loot. And the blood almost dried when her folks came home. This serpent certainly wasn't scared away from his loot by their return. To me it sounds more like a creeper, a prowler. How about windows?"

"Everything looks tight. Mac, sound out Paul Mourihan of Homicide, who's bringing you the locks. He talked to the girl's father, Dr. Clarke, about locking the door and all."

An hour later Mourihan delivered the cylinders. As I put the lock pins under my microscope Mourihan stood by, filling me in on what he knew, particularly from his interview with Dr. Clarke.

When the parents left their apartment at 8:30 to visit friends, the doctor had used his key to double-lock the mortise dead bolt on the door. He had heard Susan, his daughter, lock the Segal drop bolt from the inside. Questioned about this, he explained that he "constantly admonished" the girl to keep the door locked and not to let anyone in when he and his wife were away. To impress her with the importance of the warning Dr. Clarke always had Susan turn one of the locks from the inside herself, even though both locks had worked with the same key and he could just as easily have locked them both himself as he left.

Mourihan gave me a rundown of the shocking scene in the girl's bedroom. Immediately after the discovery, Dr. Clarke had called the police and both Mourihan and Detective Gottlieb showed up within fifteen minutes. Uniformed men were already there, trying to soothe the hysterical mother. An ambulance was on the way.

Susan, dark-blonde and slight, was lying nude on her bed, bleeding and unconscious, her face frightfully swollen and discolored from the beating she had taken. Her breasts, thighs, buttocks, and mons veneris bore numerous razor-blade cuts and cigarette burns. Semen had dried on her belly. Panties, which had been stuffed in her mouth as a gag, lay next to her face. Her pajamas and a robe were strewn on the floor.

When the ambulance arrived, Mrs. Clarke rode with her daughter to the hospital. The father, Mourihan, and Gottlieb followed in a squad car. Uniformed men stayed in the apartment to preserve the crime scene.

Mourihan's early arrival at the apartment was indicative of the girl's precarious condition. He was Homicide and Dr. Clarke's telephoned report to police left doubt as to her chances of survival. Riding with the detectives, the doctor provided some more facts.

At 11:00 that night he had phoned his daughter. Apprehensive when he failed to get an answer, the parents immediately left their friends and took a taxi home, arriving at about 11:20. Dr. Clarke recalled that when he used his key on the Segal lock at the top he found it already unlocked and he had only to give the lower lock a one-quarter turn with his key to open the door.

"I had a terrible feeling," he said. "I dreaded going into our apartment. And when we saw our child . . . That man must be some kind of savage animal!"

As Mourihan wound up his story for me, my microscope was verifying my hunch that the crime was not the work of a pick man. The lower pins of the lock cylinder were unmarred; none bore the track of a pick. The marks made by such burglar's tools as a picklock, rake, or snapper (a vibrating pick) are distinctive. Not only are they peculiar as to form (dents, gouges, scratches, and striations) but the various picking instruments bruise the pins in a manner, and in places, that a key cannot. An expert can determine not only the specific type of tool used but can come up with a very close estimate as to how much time the pick man needed to defeat the lock, as well as his degree of skill and experience.

"Paul," I said, "it's no good. Those cylinders were not picked. What about the keys?"

"Nothing there that we can see, Mac. They had three keys, one for each member of the family, and all are accounted for."

"Does the building require a key for emergency?"

"It does but the doctor just wouldn't give them one. He really worried about his place being secure."

"O.K. How about this: We know the door was double-locked at eight-thirty, but what if the kid went out for a few minutes, came back in, and only slammed the door? Leaving it single-locked. Only the snap lock would be holding it. Then the guy could use a celluloid strip to open it and show no marks."

Mourihan was unconvinced and we kept kicking ideas around. "The girl," he reminded me, "was in her pajamas when her parents left. She was found nude, with the pajamas on the floor, when they came back. We doubt if she went out but maybe somebody came to the door, a neighbor or someone else, and she opened it to talk and then left it single-locked. I guess that's possible."

"You or Frank can check that pretty easily. But if she did not open the door to a neighbor, then Frank's thought about the girl letting the rapist in may be right. At least that's the direction you'll have to take for now."

"Listen, Mac, the girl was pretty well trained about the locks. If they weren't picked, well, that's it. We're going to have to ask her mother and father all those lousy questions about their daughter's friends. And boy, she's a pretty thing, too."

I couldn't help feeling queasy about what would happen.

"It's a shame," I said. "Her school friends will have to be asked about her reputation, where she hangs out, with whom, and all the rest of it. Fifteen years old! It can't do her any good, Paul."

Mourihan agreed. "The thinking has to be, what kind of girl would let a creep like that into the house with her folks out? A guy who could do such a thing. It's going to be a real mess."

"If she surfaces she'll tell who it is, and we'll be able to keep it quiet. *If* she surfaces."

I scratched my initials into the cylinders for future court identification, if necessary, and typed a brief report of my findings. I gave Mourihan a copy for Frank Gottlieb.

Late that afternoon Susan regained consciousness at the hospital. She was asked one question by the policewoman who sat patiently at her bedside:

"Who did this to you?"

Susan whispered: "I was in my room doing homework and had the radio playing softly. I heard a noise and there was a man in my room, standing there."

The policewoman asked her nothing more, then telephoned the statement to the precinct. When the detectives there heard of the conversation they were stunned. What the hell was this? She had to know the guy, she let him into the apartment, didn't she? Didn't she?

Shortly after midnight my home phone rang. It was Deputy Chief Inspector William Kimmons, Commander of Brooklyn detectives.

"Mac," he wanted to know, "how sure are you about those locks you looked at today in that Brooklyn rape?"

My mind went back to the Bausch & Lomb magnifiers and I could see those pins clearly.

"I'm positive," I said. "This was no pick man. Why?"

I could barely hear his answer: ". . . because something's not right. The girl is back with the living and she says the guy just appeared in her room. Out of nowhere."

I stood my ground. "It sure sounds like a pick but I'm telling you, it's not."

"If she let the man in like you say, Mac, she sure as hell would not be protecting him now. This was damn near a homicide and the kid's still not out of the woods."

"Hold it, Chief," I protested. "I didn't say she let the guy in. I told Brooklyn that the cylinders given me were not picked—and they weren't. I was also told the place was tight. Maybe the girl did open the door for some reason after her parents left."

"To let the guy in? I told you, Mac, we believe her."

"Not to let just anybody in, maybe a girl friend came to the door. The kid slams it and it's not double-locked, and a 'loid is used. But if we believe all we've been told, the girl was warned repeatedly about keeping the Segal on—"

"I don't know, Mac. Look, I'm here now at the apartment. How long will it take you to get here?"

"About an hour, Chief. Give me an hour."

Driving to Brooklyn I kept thinking how awful it would be if I was wrong. That girl, not only beaten and raped, but having the police and her friends believe she had invited it. Yet I was sure those pins hadn't a scratch on them.

The apartment was on the seventh floor of a well-kept building. Two shiny new cylinders, advertised as "pick-resistant," had been installed in the apartment door. Detective Gottlieb led me into the kitchen, where Chief Kimmons sat opposite a graying, unshaven man who struggled wearily to his feet as I was introduced.

"Mac," said Chief Kimmons, "this is Doctor Clarke. Sit down, the coffee's fresh. Let's talk about this."

I sat—and found myself looking directly at a door. Another door? I couldn't believe it.

"Chief," I said, trying to control my voice, "what's this door? What about the lock on it?"

Before Kimmons could answer, the doctor volunteered: "I spoke with some other detectives about this kitchen door. When we came home last night the chain was on just as it is now and I told them I know it was also on when we left our daughter."

The chain bolt lock sat in its slide, secure for all the world to see.

"Mac, you didn't know about this door?" Chief Kimmons was squinting at me. For him to ask that cutting question in the presence of an outsider signaled an upcoming explosion.

"No," I said. "I did not know about another door."

Kimmons's jaw tightened. Staring into his coffee cup, he slowly wrapped his trembling hands around it.

"All right, you know about it now. *Both* of us know about it. Take a look."

It was a metal-clad (calomine) door leading to an interior fire stairway held shut only by a rim-mounted spring latch. Dr. Clarke, now acutely alert, offered: "I know those locks aren't much good so I had the chain put on for better security."

I stepped to the door. About six inches from the end of the channel in which the chain slid, and level with it, I could detect and feel a residue of sticky substance. No marks were visible on the beveled edge of the door's spring latch but there seldom are when a celluloid strip is used on a lock.

The sticky stuff on the metal door, independent of everything else I now knew about this case, was for me conclusive evidence. Yes, our savage rapist was a burglar. Applying a piece of celluloid to depress the vulnerable snap lock, he had opened the door only to be confronted by the chain bolt. He reached through the opening and attached a rubber band to the inside of the door just beyond the end of the slide channel. To hold the band, he applied masking tape. Which left the surface tacky.

It was now clear that the girl had not admitted her assailant. We were looking for an experienced, semiskilled burglar, one who had put the chain back in its slide after entering here and who had left by the front door. I explained what must have happened.

Chief Kimmons, peering over my shoulder, showed a great relief at this development.

"Thank God," he said, "we got to this early in the game."

Out of the father's hearing I whispered, "What about the girl? The investigation into her habits, her friends?"

"No, she's O.K., Mac. I called a halt earlier, when I heard she had said it was a stranger. I just didn't think she would cover for him. The only questions being asked have to do with any weirdos hanging around the school or this building, if the girl told anyone she had been followed, any complaints about obscene phone calls, and that kind of thing. I assured the doctor of that before you got here. Now you can have your coffee.

"Mac," he went on, "I've heard about the rubber band and the thumbtack for a wooden door. I never knew about the masking tape for a metal door. But that's beside the point. My detective thought he did know everything—he *assumed* this door was tight."

Four days after the Susan Clarke rape occurred, detectives specially assigned to the case were sent back to their own commands. This left Frank Gottlieb charged with the "open" (unsolved) investigation. Frank put whatever time he could spare into it, following up on sex crime reports and arrest messages coming over the teletype. Little else could be regarded as likely to produce results. Susan was rapidly regaining her health (physically, at least) and the Homicide men went on to other business.

About two months later, a thief in his mid-twenties was busted while fleeing from a robbery less than a half-mile from the rape scene. In broad daylight he had followed an elderly woman into the elevator of an apartment building. Holding a knife to her neck, he had grabbed a wallet from her purse. Putting the knife and wallet in his pocket, he then pummeled her with his fists—"to slow her down," he later told detectives. Running from the building, he tumbled headlong into the arms of a cruising radio-car team.

Detective Frank Gottlieb, reading the arrest message on the teletype, was struck with the wanton nature of the attack and its proximity to "his" rape case. Frank does not buy the theory that burglars never steal with violence. So he and his partner sought permission of the Brooklyn District Attorney's office to talk to the prisoner, Gil Lasen, about the attack on Susan Clarke.

A rangy, curly-haired six-footer, Lasen spoke softly, moved like a cat. Now and then, we learned, he had worked as a rug installer. His "yellow sheet" (criminal record) listed four previous arrests, significantly all for burglary. He had twice been convicted and there was no note of involvement in sex crimes or robbery.

Curious to see what the police had on their minds, Lasen had agreed to the questioning. When the Clarke case was brought up he put on a naive act: "What's it all about?"

"You can have your lawyer if you want to," Gottlieb suggested. Lasen, cocky, shrugged and insisted, "I didn't have anything to do with any rape."

Then Gottlieb decided to take the "logic" approach, appealing to the man's survival instincts. It went like this:

"We know you went there to kill that kid. It was deliberate. It was premeditated. Help yourself, Gil. If you make us go to bat with you—go to trial—you'll get the limit. It means a sentence for attempted murder, besides rape and burglary.

"Gil, you know about forensics? The labs? We have skin scrapings from the girl's fingernails. We know your blood type, too. If you were wearing gloves when you went in, you took them off when you raped the kid—your fingerprints! Do you know that semen is identifiable? What do you think led us to you, why would we be talking to you about this thing when you're in here on something entirely different? Gil, you *know* we have it. Be smart, help yourself. . . ."

At first stonily silent, Lasen finally opened up:

"I went up the fire stairs because I know they have snap locks on these fire doors and I use a 'loid strip on my jobs. All the way up I listened at doors and when I came to this one, about the seventh or eighth floor, I didn't hear any noise inside. I had no idea who lived there. I was just looking for a score, that's all.

"I used the 'loid and when the door opened I saw the chain was on. I could hear music very faint, and like somebody humming and I guessed, I had a feeling, it was a girl. I figured I'd just sneak in and take a look.

"I slipped the chain off and then closed the door behind me. I left the chain off in case I had to get out in a hurry. I walked into the place until I could see this girl reading in the bedroom. She was alone. I thought I'd fool around with her. . . . I went back into the kitchen and put the chain back on the door. . . . When I left the place I went out the front door."

The victim, Susan, had proved uncooperative. Lasen threatened her with a knife, beat her with his fists. "She fainted. . . ." While she was unconscious he raped her, cut her with a double-edged razor blade from the bathroom, burned her with a lighted cigarette. He remained in the apartment only about half an hour.

At his arraignment the judge sent Lasen for psychiatric evaluation to Kings County Hospital, where he was adjudged sane. His guilty plea accepted by the court, the burglar-rapist was sentenced to state prison for seven and a half to fifteen years.

Wouldn't psychologists have classified this criminal—before the rape—as a "professional burglar"? And therefore "passive and nonaggressive"?

Rapist Gil Lasen would seem to exemplify Clinton Duffy's "sex factor" and it may be that sex is a subconscious common denominator of burglars. If that's true, then in many of the burglars McDermott has known it is so repressed as to never have been expressed overtly.

Five-time loser Lasen was young, only twenty-six, yet the generalization persists that career burglars are "men of mature years." In this context, psychiatrist Walter Bromberg maintained that these men are "never outspokenly aggressive, even when apprehended." That's just too pat.

Consider "Dummy" Taylor, a lifelong burglar killed in a shootout with Safe and Loft detectives after he committed an armed robbery in lower Manhattan. In his sixties when apprehended, Dummy pulled a gun and attracted a bullet to his heart. Too, there was veteran burglar Tony Latches. Emerging from an apartment he had just "hit," he encountered the obviously pregnant tenant as she was about to enter. Tony, a gentleman burglar at sixty-three, ungallantly threw her down the stairs.

Of course, it would be convenient to neatly categorize burglars psychologically. But McDermott, in decades of working as a specialist in the field, finds they can't all fit into psychosocial slots. To wit:

Some burglars come from broken homes, many do not.
Most are heterosexual, a few are not.
Some are violent, some are not.
Many come from poor families, some do not.
Some are well educated, most are not.
Most spend money freely, some don't.
Some are happily married, others are not.
Few are religious, most are not.
Some have a good sense of humor, some don't.
Most are racially prejudiced, some are not.
Some like the police, most hate them.
Some take an arrest fatalistically, others do not.

Certain elements in the crime of burglary, taken together, may add up to a compelling motivation. There's the challenge of the lock, the excitement, the ego fulfillment, the attractive money rewards. For the skilled practitioner, there is the ease with which he commits the crime.

Watch the master burglar at work. See him smile and charm the doorman as he strides into the lobby. See him chat amiably with the woman waiting for the elevator that will take him up to *your* apartment. Audacious, supremely confident, he has presence, *chutzpah*, gall. Wouldn't you say he is bold? That burglary is the boldest profession?

From the savage burglary-rape case of Susan Clarke some lessons can be clearly drawn: Where the wrong devices are used you risk far more than the loss of valued property. The simple snap lock on the kitchen door was easy to 'loid. The Clarkes believed a chain bolt would beef up a door's security but failed to realize it can still be quietly removed. Such oversights made the near-homicide possible.

Burglars are never gentlemen, always potentially dangerous. The rapist was capable of only rudimentary burglary. If he had encountered a dead bolt on the kitchen door when he first tried his celluloid strip he would have moved on. And Susan would have grown up unscarred.

Chapter 3 Safeguarding your private home

"IT'S NO TRICK to make it in the suburbs—that's where the money is," a veteran burglar once boasted.

For their annual eye examination at an ophthalmologist's office in town, Thelma and Mort Weisinger were gone less than an hour. Returning to their comfortable Long Island home they unlocked their front door and stepped into the foyer.

One shocked glance and the Weisingers knew they'd had an unwelcome visitor. The floor in front of the hallway closet was piled with coats and jackets, the pockets turned inside out. The living room, bedrooms, and kitchen looked as if a hurricane had struck. In the kitchen even the refrigerator's ice-cube trays had been searched. (Savvy burglars know that many housewives believe their rings are safe when frozen in ice cubes.) Bedrooms were so littered they resembled a grabbed-over rummage sale.

Among the missing valuables: Mrs. Weisinger's charm bracelet, a wallet containing $305, her mother's engagement ring, her husband's $600 stashed in a desk drawer, several bank passbooks, thirty-five Consolidated Edison bonds valued at $22,400, a portable color TV set, powerful binoculars, two gold-plated ballpoint pens, some bottles of expensive perfume, Mr. Weisinger's collection of rare astrology stamps and five twenty-dollar gold pieces secreted in a golf bag.

On a living-room sofa the couple found a bulging pillowcase (taken from their own linen closet) stuffed with a digital computer, a movie camera, a transistor radio, and a Polaroid camera. When police arrived, the abandoned pillowcase loot was explained.

"When those guys heard your car return," said a detective, "they must have gotten panicky and beat it out through your rear door. They didn't have time to take this loot.

"There were two of them," he added, picking up four argyle socks from the floor. "They took these socks from your bedroom to use as gloves and avoid fingerprints. Carrying their own gloves would be incriminating evidence if they were caught about to break in."

Mort Weisinger, a well-known magazine writer and novelist, described the couple's reactions to the invasion:

"First it was disbelief. How could this have happened while we were away for so short a time? Then I felt a surge of fear. Suppose we had interrupted the burglars and they had a gun or knife? It was nerve-racking. Analyzing our emotions later, we thanked God we were alive. Then came a deep, uncontrollable sense of outrage. Our privacy had been violated!"

Thelma Weisinger recalls that she was engulfed by after-shock waves and simply broke down. "I don't want to live in a house anymore," she sobbed. "Mort, what will I do when you go out of town for days on assignment? Let's move to an apartment that has day and night security guards."

The Weisingers still occupy their house. It is now protected by what Mort calls "a fail-safe" burglar alarm. And all their remaining compact valuables have been deposited in their bank vault.

How did those break-in pros make their entry? This was the theory Mr. Weisinger offered the police:

"That morning we had a plumber come in to shut off the water in our swimming-pool cabana, to prevent the pipes from freezing. He brought along an 'apprentice' who looked like a long-haired hippie. When we got home after the burglary we found the kitchen door leading to the pool had not been locked. So we figured that when the plumbers left through that door someone pressed the button that would keep the door unlocked.

"After checking out the alibi of the plumber's apprentice the police couldn't hold him. But, they said, he could have tipped off a contact, a confederate, to tell him the door had been left open and the way was clear. The police also told me one technique or strategy was for another member of the gang to keep circling my house in a car during the burglary, ready to pick up the burglars and the loot as soon as they left our house."

That is safer than parking in front of the house or in the driveway. It's also not unusual for house burglars to use walkie-talkies.

Despite the Weisingers' property losses and sense of outrage, they were far luckier than Mrs. Sheila Watson and her close friend, Mrs. Bonnie Minter. Shortly before 5:00 P.M., a light-blue van drove up to the Watsons' $150,000 home in Lewisboro, New York, near the Connecticut border. Two men got out and walked into the house, surprising Mrs. Watson and two small children.

A few minutes later Mrs. Minter arrived with two little boys. All were herded upstairs, the children to one bedroom, the mothers to another. The intruders shot the women to death—Mrs. Watson eight times, Mrs. Minter three times, both of them in the head, back, and chest. Then the criminals ransacked the house, carting out the loot to their van.

As Major John Leary of the State Police viewed the crime, the violence was unusual for what seemed to be professional burglars. Two days later three suspects were picked up. How were they tracked down?

Residents of the largely rural area reported noticing a blue van before and after the slayings. All night and the next day, police in New York and nearby Connecticut kept stopping blue vans on roads and highways. Then Major Leary shrewdly came up with this suggestion:

"We had an idea the blue van had been rented. Rental vehicles are often used by burglars."

And so state police were waiting at the Taylor Rental Service station in Norwalk, Connecticut, when three men drove up to return a rented blue van. Their van fit the description of the vehicle seen by the victims' neighbors and the men were subsequently arrested, two of them charged with murder.

How you are vulnerable

Safeguarding a private home presents far more of a challenge than does making an urban apartment secure. A house has more windows that are easily accessible, especially on ground level.

House hidden by hedge or shrubbery allows intruder to operate unseen.

Doors often have glass panes. Basement windows and doors, casement windows, and patio doors represent additional problems. A garage may have an inside door leading into the house. Dutch or French doors and air conditioners may readily be reached from the ground.

A tall, thick hedge or shrubbery around a house permits an intruder to operate without being seen from a road or nearby houses (see photo). An unprotected attic window may be approached with a ladder left lying around and the burglar's movements are concealed by an old tree. If, following a family's pattern, no one is home until evening and mail is left in a curbside box all day, the smart thief knows the way is clear for him.

While shopping at the supermarket, a housewife leaves her garage door open. A vacant garage, visible to all, serves to welcome a lawbreaker. Thus, a peak period for home burglars is between 9:00 and 11:00 A.M. When the family is temporarily away from home, a house key is often conveniently left under the outside doormat, in a flowerpot, or on the ledge over the front door. Experienced thieves know this.

Ranch-type homes are popular with certain burglars, especially houses with sliding doors opening onto a patio or porch. Unless modified and equipped with a superior lock, these doors can be a cinch to open. Often rear doors surrounded by greenery offer excellent cover. At a house on a large plot, especially, the noise a burglar may make without being heard allows him to smash even picture windows, if necessary.

Some people make the mistake of depending solely on an alarm system as a deterrent. Talented thief Bernard L. Foley, cruising through suburban streets at night, would look for such telltale signs as a pile of leaves in front of a garage door, the lights out in the house. Circling the house on foot, he'd listen at windows and doors for sounds of a radio or television, or a barking dog.

One night, confident it was safe for him to move in, he came upon a decal on a window advertising the presence of an alarm system. He knew that this system was of the telephone-dialing type. He simply pulled out his pocket knife and cut the phone wires where they extended from the house. Back in his car he waited for a half hour to make sure there was no response, then broke in by using his vise-grip pliers on the key-in-knob lock (see photo). (*That break-in could have been prevented if the homeowner had had a deadlock on his front door.*)

Suburban thieves

Homes in middle-class suburbs, are considered sitting ducks by this breed of burglar. There, houses are usually detached

How vise-grip pliers (center) are used on lock cylinders for forcible entry. To prevent this, use a cylinder guard ring or guard plate.

enough to allow the man (or woman) to function unobserved. Roving around, he selects his targets from what he notices: no lights in the early evening, vacant garages, other signs of welcome.

Many of these criminals commute from a nearby city, and they generally operate in pairs. Often they work more than once in the same locality. While one man rings a doorbell—and breaks in if he receives no response—his partner drives around the block until it's time to pick up the gathered plunder. They may use a panel truck or van to cart away such items as color TV sets or hi-fi equipment.

The commuter-type housebreaker sometimes first loiters around airports. There he reads the names and addresses on the luggage tags of people bound for remote destinations. Then he knows where he can burglarize, assured that the owners will not be around for quite a while. In some instances, this burglar will strip a house of furniture, rugs, and other household goods. (*To avoid such exposure, use your business address on luggage tags.*)

Daytime burglaries of houses account for over half of all residence break-ins. Most often they are committed while homeowners are at work or on vacation. During the day thieves can move around freely on the local scene and learn the habits of potential victims; and it's easy to pose as someone on a legitimate errand, such as a repairman or gardener.

Jack O'Malley, one of the most brazen daytime operators, rambles around a neighborhood on a summer weekday and sees a woman gardening in her backyard. Ringing the front-door bell and receiving no response, he tries the screen door. If it's unlocked he walks in; if locked it's likely to be "secured" by a simple catch-type device. If he must, Jack slits the screen, reaches his hand inside to the device, enters, makes his way quickly to the master bedroom, and steals everything worthwhile that he finds.

(*To discourage the intruder, fit the screen door with a more substantial lock*—either a key-controlled chain bolt or at least one that may not be reached if the screen is slit. A sliding bolt at the bottom of the door is effective. That door is usually converted in winter to a storm door; the keyed chain can then be used to interview strange callers.)

Around Wethersfield, Connecticut, at least three families recently reported they had been victims of a female burglar. Described as in her late thirties, she worked in the early afternoon, generally wielding a jimmy to force windows or doors. According to local police she was responsible for about forty burglaries within six months.

One afternoon a housewife taking a nap was awakened by odd noises on the ground floor and went to investigate. In her kitchen she plainly saw this strange woman applying a "long metal tool" to the window. When the housewife shouted at her, the uninvited visitor fled.

In a successful burglary soon after, the thief had come in through a rear window on the ground floor. The screws on the window lock had been pulled out of the wood, and on the glass-paned back door all four sides of the molding on a pane had been taken out. The homeowner had a lock on that door that could be opened only with a key from the inside, so the intruder couldn't get in this way. Yet the thief had persisted and managed to enter through the window.

Upstairs, drawers in the bedroom had been dumped on beds, contents of closets were scattered on floors, pockets of garments turned inside out. Much cash and jewelry were missing. A neighbor had noticed a "woman stranger" walking away from that house with a heavy shopping bag just minutes before the homeowner arrived home. She had been away only forty-five minutes.

Burglaries committed at night are often traced to young thieves, although daring veteran practitioners will also look for favorable conditions from the early hours of darkness until about 10:30 P.M., choosing unlighted residences. They opt for places they know to be uninhabited.

The recent trend pointing to an increase in daylight burglaries in private homes over nighttime break-ins may be attributed to a greater awareness by the citizenry of the mounting prevalence of thievery. Most people are naturally more vigilant of possible break-ins when they're away from home at night, so they bolt doors and lock accessible windows. But many of these homeowners are not so diligent when they leave the house during the day for a short trip to town or to a supermarket. Nevertheless, gangs of young burglars are usually reluctant to work in daylight; they feel too exposed and they probably lack the skills to break into a house quickly. Their crude methods of entry need the cover of darkness.

A thief working on homes can be potentially more dangerous than one operating in city apartments. If he rings a house doorbell and receives no response, the burglar will believe no one is home. But the owner may not have heard the bell. Commonly the invader, having made a forcible entry, goes in with a deadly weapon—his jimmy.

Many a bold burglar is not fazed even by the presence of the family at home. Some prefer a doctor's home, probably based on the widely held assumption that all doctors are rich and have money salted away at home.

Dr. C. T. was acutely alert to his position as a prime "mark" because of his profession. His car, parked on city streets, had been broken into several times by "idiots" looking for narcotics, so he had installed an alarm in his car as well as in his suburban home. He had a peephole and a chain bolt on his front door and the door was kept locked at all times. The alarm was turned on just before the household retired for the night and whenever the family was out.

Relaxing in his living room watching television one evening, hoping that one of his patients wouldn't disturb his rest by going into labor until next morning, the doctor heard a slight sound and caught a movement out of the corner of his eye. He turned. Two men were standing in his living room, one holding an ugly gun, the other smiling as he placed small metal instruments back into a leather case.

At first, Dr. C. T. wouldn't cooperate with the stickup men and adamantly refused to tell them where he kept cash and his wife's jewelry. He was pistol-whipped and the gun was held to his teenage son's head. The intruders had their way.

Apparently disinclined to depend on a subterfuge to get into this carefully "protected" house, the intruders had taken advantage of the doctor's one point of vulnerability: his reliance on an *ordinary* lock that was easily picked. Moreover, the doctor had failed to keep his alarm on during the early evening hours because he felt safe staying at home. The burglary had turned into the more dangerous crime, robbery.

Where and how they make entry

In private homes fewer break-ins are committed by picking of locks. When the pick man does work suburbia, he generally goes for the front door. His strong point: getting in very quickly. A neighbor who sees him ringing a front-door bell, and then a few seconds later walking in through that door, is unlikely to run for the telephone to call the police emergency number. It all appears so innocuous. The thief's talents give him a distinct advantage over his less adroit colleagues. He doesn't have to walk around to the back of the house looking for weak points.

By and large, however, pick men tend to prefer the anonymity of apartment and hotel corridors. So most suburban thefts are accomplished by the thief who forces a door or a window in the rear or on a secluded side of the house (see photo). He is most

A secluded side of a house offers a burglar the chance to force a door or window unobserved.

likely to break a pane of glass in a door. Then he has only to reach through and release the lock. Police believe this is how an intruder entered the home of Senator Charles Percy in a Chicago suburb before stabbing to death his twenty-one-year-old daughter.

Since most glass in doors or windows is easy to break, use of glass cutters has been extremely rare. When adhesive tape is applied, it muffles the noise and holds the pieces together so that the criminal can pick them out and stick his hand in without cutting himself.

Normally, when an air conditioner has been mounted in a window, the thief can merely exert a little leverage to lift the window and remove it. Air-conditioning units mounted through walls are seldom fixed firmly in place; pushing them into the room (or pulling them out) presents the intruder with a convenient opening through which to crawl. He can also walk off with the air conditioners.

The larcenist will try basement windows large enough to admit him. Too frequently such a window is left open or held shut with a flimsy device. Some burglars prefer second-story side windows facing a garage or trees because such windows are less visible to a neighbor or accidental passersby and will rarely be given security attention by homeowners. In these instances the intruder will look for a ladder in the yard or in an unlocked garage.

Then there's the common mortise-lock specialist. The thief rings your bell, you answer it, and he gives you a cock-and-bull story that persuades you to leave him alone with the door open for a minute or more. On the edge of the door where the lock is set there will be a metal plate covering the lock. In many cases, there is a screw exposed on this plate which holds the lock cylinder rigid in its place. Using a screwdriver, the thief moves this screw just one and a half or two turns to loosen the cylinder. Then he walks off and waits in his car a short distance away for the occupant to depart. He now has a setup to which he can return and unscrew the cylinder by hand.

(If you have a mortise lock, make sure it's the *modern type* with an additional plate fixed over the lock mechanism to cover the cylinder set screw hidden under the plate. See photo.)

A thief blesses the side or rear door that opens inward. If its

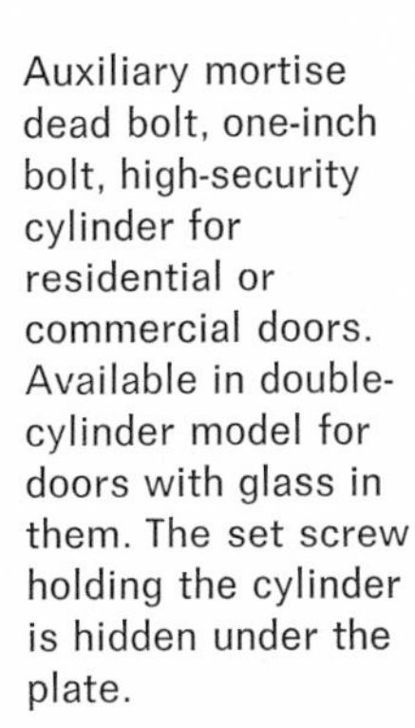
Auxiliary mortise dead bolt, one-inch bolt, high-security cylinder for residential or commercial doors. Available in double-cylinder model for doors with glass in them. The set screw holding the cylinder is hidden under the plate.

flimsy lock resists him, he simply kicks it in. Once he is in, a veteran burglar proceeds to open a back door and a window on a side away from the garage. Thus he has three avenues of escape in case anyone does come home while he's inside. Whichever entrance the homeowner heads for, the burglar avoids, choosing another exit.

In recent years many homeowners have been attracted to lock cylinders for their doors which require a round key (see illustration on p. 60). This type of key has been common in residential alarm systems. The homeowner uses such a key to arm or disarm his system by placing it in the lock mounted on the doorframe, and then employing the same key to open his door.

"Fat Albie" Hirsch developed an ingenious method of burglarizing homeowners who rely on these devices. He would go from house to house posing as a solicitor of an orphanage. In his coat pocket he carried a can of hair spray. As opportunity presented itself, Fat Albie would direct the lacquer spray at the lock, saturating the circular opening. Then he'd sit in his car, parked far enough away so that he wouldn't be readily seen, yet able to watch for any departing housewives.

While he waited, Albie knew the hair spray would be doing its job—getting tacky and a little stiff. Finally, his vigil was re-

Lock requiring a round key. This photo shows a burglar's tool (picklock) to defeat the lock. At bottom right, the round key.

warded. A woman emerged from her front door and locked it, used the key to set the house alarm system, and confidently drove off. Minutes later, an equally confident Albie strolled up to the door. Simply by turning the round keyway in the alarm, and then doing the same with the door lock, he walked in.

Hirsch, having too often encountered this lock, had experimented with it at his kitchen table. Incapable of picking locks, he found that if he sprayed the face and keyhole with lacquer and used his key in it a short time later, the hair spray would make the locking pins stay in an "open" position. He could then rotate the cylinder, using anything at all. He was in business.

Summer homes

Vacation homes in rural and resort areas have become happy hunting grounds for burglars as well as vandals during off-season. These scattered cottages, cabins, and bungalows—at lakes or in the mountains—cannot, of course, be patrolled well by the typical local police force. It's not at all unusual for outlying vacation houses to be completely stripped of their furniture.

Often women who normally have a man around the house are alone during the summer when the family takes up its vacation

Above Doors of summer homes invite attack by burglars.

Below Inside of the door of a summer cottage that some thief considered vulnerable.

residence. Except for weekends, the husbands generally remain in the city. The summerhouses are probably fairly isolated (see illustrations). Yet there's a tendency for owners, particularly city dwellers, to relax their guard.

If you've a hideaway house, take the first of those nice long weekends to insure against break-ins by paying attention to the basics of door and window security.

Vandals and youthful thieves are your chief concern in a summer retreat unoccupied all winter. Even if you disconnect the electricity, protect your property by installing a battery-operated alarm system. If your place is so isolated from neighbors that the sounding of a bell or siren will not be effective in warding off burglars, investigate the possibility of a police-connected alarm. In many small communities the police will accept a telephone alarm line into their headquarters. (For more on alarm systems, see Chapter 11.)

The greatest threat

In the context of burglary, lock picking represents the greatest menace to your personal safety. Listen to what happened to a woman in Florida who was convinced she was perfectly safe behind locked doors. Says McDermott:

I once collared a pick man who realized he had to plead guilty. Gus opened up to me on a strange series of events in the hope that I'd let the judge and probation people know he cooperated.

Some months before, Gus told me, he had been given a contract through a fence to fly to Florida, rent a car, and meet a stranger near West Palm Beach. This fellow would accompany him to a private home where Gus's task was simply to pick open a front-door lock and another door on the second floor.

The owner of the house was a garment manufacturer whose wife spent her winters there alone while he shacked up with one of his models in New York. Since the man's relationship with his mistress was an open secret, his wife's murder would obviously put suspicion on him—unless the wife, who had a history of heart disease, died of "natural causes."

Gus was given travel money, a $300 advance, and assurance

that the locks he would have to open would not challenge his skills. In Florida he was driven to the house by "a creepy-looking racetrack type." His driver started to move up the long curving driveway off the quiet street, then stopped. Seated in the car, after a few minutes of staring at the darkened house at the head of the driveway, the man suddenly announced to Gus: "It's all off. I'll take you back to your car and you'd better go on back to New York." No explanation.

Several days after Gus returned, his fence—the man who set him up on the deal—told him: "You're well out of that job. The house you were to open up belongs to a guy who wants his wife killed. The contact you met either got cold feet or didn't like your looks. He was going to do the job on the woman, using her bed pillow."

After protesting that he had never before been involved in "such a thing," Gus spilled the rest to me: "I've since heard that the garment man's wife died." He did not know the businessman's name and I agreed it would be awkward for him to ask his fence for it.

I checked with the police department in towns adjacent to West Palm Beach. There were some false starts until we found a record that revealed a "death from natural causes—heart failure" for a New York garment man's wife—three weeks after Gus's visit. The police report stated that a domestic had found the woman's bedroom door locked one morning and was unable to rouse her. The woman's car was in the garage. When the bedroom door was forced open she was discovered lying on her bed, dead.

Significantly, the one key known to exist for that bedroom door was in her purse on a closet shelf in her locked room. Prescription bottles lay on the night table.

When I asked to have the lock cylinder removed and sent to me, the Florida police informed me that the house had since been sold, and the new owner had put a new lock on the front door and removed the bedroom door lock as "unnecessary." So we had reached a dead end.

A nervous woman had sat alone in her bedroom with a false sense of security, knowing she had locked her doors. As I visualized the crime, a pick man had opened the doors to let the killer in. And probably a pillow was pressed over her face, lifted, then pressed again until the woman succumbed not to suffocation but to a strained heart—"heart failure." When the act was accomplished, the lock cylinder—left in a picked position by the pick man—had only to be rotated to lock the doors, thus obscuring the murder.

We had no hard evidence on which to move. Not long after, the garment man married his model.

If the victim had used pick-resistant locks, she could have made a pick man's trip to Florida a waste of time—if she had realized how easily conventional locks are picked.

PRECAUTIONS TO TAKE

► *Study your doors.* If you have vulnerable glass in a door or in a wall next to a door, you need double-cylinder locks that can be opened only with an inside key. For extra protection, put in a sliding "barrel bolt" near the bottom of the door which an intruder can't reach.

► Before you invest in new locks, make sure your front door is solidly constructed and pay special attention to your door panels—they should not be weak. To be secure, doors should fit the frames neatly. On metal-clad doors, to resist a jimmy you need a good vertical drop-bolt lock to supplement your primary lock. On a wooden door use a horizontal dead bolt that extends one full inch into the frame. If you have a strong door with a weak frame get a police-type lock consisting of a bar angled from the door to the floor. (For more on the "police" lock, see Chapter 11.)

► The door leading from your basement to the house interior should be secured from the inside, preferably with a jimmy-resistant vertical bolt or a heavy-duty sliding barrel bolt. Is there an inside door leading from the garage to your basement or kitchen? It should be treated as an exterior door.

► As a homeowner you're free to use any locks you like. For convenience it's a good idea to have all those on exterior doors keyed alike; that is, one key operates them all, including locks on your garage and tool shed.

► Your front door should have a peephole or viewer. A glass lens inside the tube allows for a 180°-to-200° wide-angle view, so you can see a great deal; this is especially good if more than one stranger is outside the normal range of view. Also install a chain bolt to interview callers or to transact business.

► For sliding-glass patio doors, install a "Charlie bar," a key-controlled device that permits you to open the door slightly for ventilation and to lock it in that position. Locks made for narrow stile doors may also be placed on your sliding doors.

► *Study all your windows.* Those most vulnerable are obviously those on the ground level; upper-floor windows near a trellis, just above a garage roof, at cellars, and above a porch are also subject to break-ins. All accessible windows, especially those on the ground level, should be equipped with key-controlled locks and be pinned, as described in Chapter 2. Key-controlled locks and pinning will permit opening a window slightly for ventilation but too narrowly to permit entry (limit the opening to four inches).

► Basement windows, if you occasionally need ventilation there, may be secured by eyebolts and padlocks, grills, or bars.

► Most private homes have basement windows that are hinged at the top. In a crime-ridden area excellent security can be achieved with a single steel bar extending horizontally across the entire window. This bar slips into a piece of angle iron bolted to the frame at either end, secured by a padlock. (See illustration.)

► Casement windows, which offer better basic protection than the wood double-hung varieties, are generally more difficult to beef up with additional security. A burglar breaking a glass pane will reach through and crank the window open or lift the clasp handle holding the window shut. However, if he reaches through and you have *removed the crank handle* or secured the clasp handle with the keyed devices available, you've beaten him. On vulnerable windows, remove the pin that fixes the handle to the cranking shaft and simply lift the handles off their shafts when you are going out. Another alternative is to drill the body of the clasp handle to accept a padlock or removable pin.

Also for casement windows: There's a keyed lock-and-lever replacement for the standard lever action; the lock placed opposite the hinged side will prevent the window from being pried open.

► On sliding windows (and doors) you'll get extra protection

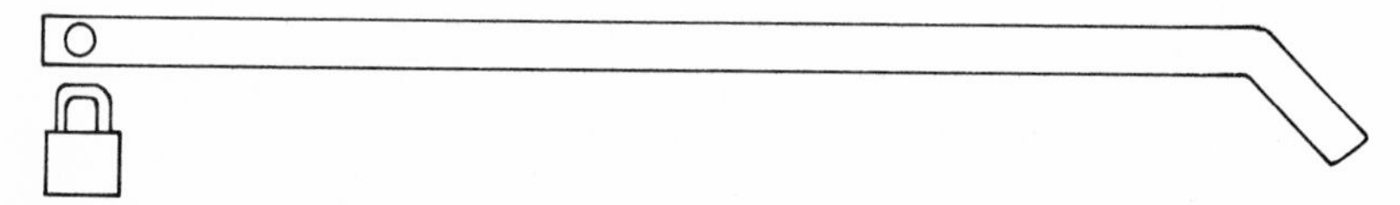

For better security: basement-window steel bar and padlock.

Above Angle iron is screwed to window frame.

Below Through the angle iron—eye bolts—insert steel rod with hole drilled for padlock.

by wedging a broomstick, cut to the needed length, in the bottom track. This should prevent anyone from opening it from the outside.

► With all window locks, every member of your family should be able to open them. If a fire breaks out it may be necessary to leave by a window. Order the locks keyed alike. Keep a key near the window in every room.

► At hardware and five-and-dime stores, you can buy all kinds of gadgets designed to wedge your windows shut or hold them partly open for night air. Most of them are virtually useless for your primary purpose: to ward off entry by a burglar. Steer away from so-called "keyless locks."

► *Consider your garage.* If it has glass windows, make them opaque or cover them on the inside with dark curtains. Since a criminal may hide in the garage, it should be lighted at night on each side of the driveway with at least a 100-watt bulb. For an open carport, one fixture suspended from the rafters usually produces enough illumination.

► Most overhead garage doors are fitted with a T-type handle that locks them. This handle is adequate if the doors do not provide access to the inside of your house. Hasps and padlocks mounted on the outside of the door are not only unsightly but vulnerable.

► A remote-control automatic garage door opener acts as a deterrent to burglars and also limits the time you are outside, unprotected. Thus, a woman driver arriving home late at night does not have to get out of her car to unlock and open the garage door. Coming up the driveway she actuates the lock, drives right in, and it closes automatically. With such a device, if you are going away on vacation and your garage provides access to your house, it would be worthwhile to secure the door from within. Insert a padlock through the roller track just above one of the rollers and then leave through the house.

► Lighting threatens intruders by exposing them. At night, keep the breezeway from the garage to the house well lighted. If yours is a two-story house, leave a light on upstairs as well as downstairs facing the street but draw your shades so that unoccupied rooms can't be seen. On the exterior, a light should be on all night at the front door as well as at the rear and porch. Install sunlight-sensing controls in each lamp; these turn lights on at dusk and off at dawn. An inside light kept burning all night can be muted by a drawn shade or venetian blind.

► One good way to leave lights on is to buy a timer that operates at random periods. With one type, General Electric's variable timer, a single setting will turn a light on at 6:30 P.M. and off at 11:30 P.M. one night; the following night it goes on at

7:00 P.M. and off at 11:00 P.M. Then it recycles. Another sophisticated timer, Paragon's Touch Command, turns the same light on and off a few times during a twenty-four-hour cycle.

► Do you have a shadowy garden? Have a light-mounted post installed at enough distance from your house so that it floods any shrubbery offering concealment or otherwise dark corners of the house exterior. (Of course, the lights should not shine through your bedroom window or your neighbors'.)

► The best light sources are incandescent lamps and deluxe white mercury lamps, which are more efficient and have a longer life span.

► If you have air conditioners in windows accessible from the outside, bolt them to the supporting frame underneath. The window should come down snugly on top of the unit. Be sure the window is immobilized. Pin it as described earlier (see Chapter 2), and block the window track with a length of dowling (instead of the small angle piece, as instructed by the manufacturer, to screw to the top of the window in the frame). This should prevent the window from being moved up to remove the unit. For permanently mounted wall units, secure the air conditioners to the inner wall by bolting a piece of metal to the sides or bottom of the unit, and applying screws to the wall studs.

► In hot weather, a thief will know if you're home by whether your air conditioning is operating. So when you're out, keep the unit running at low speed.

► *For outer defense,* a wrought-iron, woven-wire, or picket fence with a gate can be a psychological deterrent, provided it is not so low that the fence can be stepped over easily. Locked gates and fences also deter removal of large items. Your gate may, in addition, have an alarm attachment against night prowlers. The fence should not present a hazard to show-off children.

► Lock all ladders in your garage or your basement. Remember, the Lindbergh kidnapping was accomplished after Hauptmann used a ladder to an upper bedroom window.

► Have you lost your keys? Change the cylinders on all locks.

► Keep the phone number of your local police department at your phone and at all extensions, particularly in the master bedroom.

► If you put your house up for sale insist that all prospective buyers be accompanied by your real-estate agent.

► Do not display your name outside your house. It enables a cruising burglar to telephone you and assure himself of your absence.

What about burglar alarms?

While no one device will solve all your problems, a *good* alarm system is the single most effective precaution. In a private home, a decal advertising the presence of an alarm system that by reputation thieves have learned to respect can convince them to look elsewhere for a victim. Anything less will attract a Bernard L. Foley.

You probably need an alarm system if: (1) you live in a section where many burglaries have occurred and you have trouble getting insurance; (2) you own fine art, a valuable coin or stamp collection, rare books, and/or expensive jewelry; (3) your house is isolated from neighbors; (4) you live alone and are scared of physical attack by an intruder. Alarm systems can seem both expensive and inconvenient, but you may save on burglary insurance premiums; and a good system can save your life.

The expense of an alarm system can be reduced by limiting its perimeters—for example, excluding such easily penetrated spaces as a sun porch. Or, instead of wiring the entire house, you might install a space-protective sensor, such as one ultrasonic motion detector, in the room through which an intruder is most likely to pass.

There are two basic kinds of alarms: (1) a local device designed to frighten off intruders, and (2) a silent alarm that signals police headquarters or a security service from which guards may be dispatched. The least costly alarms are "contact" devices, mechanical switches attached to doors or windows. The contacts form an electrical circuit that sounds an alarm when broken.

A suburban Chicago neighborhood had been plagued by burglaries not long ago. A twice-victimized homeowner, going away for weekends, decided to fortify his house as if it were a Fort Knox. Late one Saturday night a pick man, opening a door lock with his small steel tools, ventured into the darkened living room. All hell broke loose. Sirens shrieked, bells clanged, lights went on and off, a revolving red spotlight on the roof flashed like an airport beacon.

Scared out of his wits, the burglar dove through a window, badly sprained an ankle as he tripped on the lawn, and lay dazed and babbling when cops nabbed him soon after. An electronic alarm linked to the local police station had tipped them off. Of course, you can overdo your alarm fortification. Be selective. For specific choices, see Chapter 11.

Where to keep your valuables

Put your stocks and bonds in a bank safe-deposit box or have them retained at your broker's office. Compact valuables should also be placed in the bank's vault. Generally you don't have to buy an elaborate household safe; burglars find them and have been known to cart away safes weighing as much as a thousand pounds. The main reason you may need a safe is to protect papers that you refer to often. It must have a fire- and heat-resistant lining to protect documents from fire.

Money chests are more burglarproof, especially when bolted inside a safe. You might conceal a small chest in some place unlikely to be approached by a thief, such as the attic. A closet floor is also a convenient location for a chest, provided it's not in the master bedroom.

Don't leave valuable articles in your master bedroom; that's the first place a burglar heads for.

A so-called "security closet," fitted with a solid dead-bolt lock, may not be impregnable. In fact, a determined thief may view it as a challenge. Aware that a locked closet must contain valuables, he may break out a wall to get inside.

Ordinary key-locked or combination-lock file cabinets, chiefly to preserve papers against flames, are easy to break open—usually with a jimmy or by inserting a long, thin knife blade along a drawer top and pressing it against the locking bar inside.

Adding a steel bar that slides through all handles, and is locked by padlock and hasp, will help if the padlock is a good one.

Avoid blurting out where you've hidden your treasures. Ponder the tragicomic case of a citizen we'll call Mr. Lawrence, proud collector of precious rare stamps that he kept in a safe. When a New Year's party at his home started to drag, he suggested playing a guessing game. Half-soused, he offered a bottle of twenty-five-year-old Scotch to any guest who could find his safe. The safe was concealed in his basement, a "finished" room.

After a vain search, his guests gave up and prevailed upon him to prove the safe existed. Triumphantly, Mr. Lawrence showed them how he had built a false flight of stairs under the exposed stairway leading upstairs from the basement. Into one of the hidden steps a small safe had been cleverly installed. The guests cheered.

A week later Mr. Lawrence's safe vanished. It turned out that two of the guests at that party were also rare-stamp collectors. Detective Robert McDermott, called in on the "no forcible entry" theft, guessed that one of the two "friends" must have commissioned a pick man to do the burglary. It couldn't be proved until about a year later when a crack burglar was arrested on another job; to help himself out of the spot, he told the story to McDermott.

Stashing cash away in a refrigerator, taped to the bottom of a dresser drawer or in your toilet tank, in a hatbox on a closet shelf, or in clothes—these stratagems are familiar to experienced thieves. A better idea is hiding valuables in a child's room (where the child can't reach them), because burglars generally bypass rooms full of toys.

A well-known mystery-story writer regarded himself as remarkably astute at outwitting any potential burglar. Leaving his home on a long weekend, he placed his wife's jewelry and other small expensive items in an ordinary paper bag and "hid" it on top of the water tank behind his bathroom toilet. What burglar, he figured, would look in a bathroom? He had to be told—too late—that the average burglar, tense during his activities, is apt to use the toilet. So he'd be standing staring at that bag for maybe half a minute before he just reached and opened it.

PRECAUTIONS TO TAKE

When you're out for the day or evening

► Give your house a lived-in look. Away for the day, don't allow packages to be delivered to your door. Inform a dry cleaner or department store to deliver to a neighbor's because you're confined to your home with the flu.

► If you must leave your key somewhere, rely only on a trusted neighbor.

► Planning to be away all evening? Leave a light or two on, close your blinds, draw the curtains, and pull down shades so that a thief can't see into a room. If you tilt the slats of blinds upward he can see only the ceiling. With a light on in a bathroom, leave the door open a crack where the light shines outward. Mess up a bed. If a burglar is bold enough to break in despite the light showing, he'll see the disturbed bed and the bathroom door open partway. Usually he'll exit quickly.

► Never leave your garage doors open when you go shopping.

► Here's a tongue-in-cheek suggestion by Woody Allen: "What's the first thing the average homeowner should do to protect himself against burglars? When you go out, keep a light on in the house. It must be at least a sixty-watt bulb; anything less and the burglar will ransack the house, out of contempt for the wattage."

Before you leave on a trip

► Avoid having your home appear to be unoccupied. Of course, stop mail and milk deliveries. If you will be away for a few days, replace a mailbox that hangs outside your home with a mail slot in your front door that enables the mail to fall inside the house. In rural areas have a neighbor bring your mail in.

► Have your lawn mowed and watered regularly.

► Don't advertise a vacation by publicizing your itinerary beforehand in the local newspaper. Don't ask a service station mechanic to be specially diligent in servicing your car because you're planning a long trip.

► If you're not using your car on the trip, park it in your driveway.

► In winter, arrange to have a high school student shovel the snow at your entrance and on your driveway.

► If you will be off on an extended vacation, notify local police to keep an eye on your house.

► Do you subscribe to a phone-answering service? Tell the supervisor not to inform callers how long you'll be gone.

► If possible, get a friend or relative to serve as your "house-sitter," or at least visit occasionally. Be sure to introduce him or her to your neighbors. If your house has a burglar alarm, explain its operation.

► Without a house-sitter, have a neighbor pick up throwaway advertising sheets so that they don't accumulate.

Be a good neighbor

When we asked Milton G. Rector, president of the National Council on Crime and Delinquency, for his best advice on home protection, this is what he advocated:

"One of the most effective ways to protect one's home or apartment is the one most commonly neglected. It is simply taking the time and trouble to be neighborly and thus encouraging others to be neighborly, too.

"Part of neighborliness is keeping an eye on the house next door and taking the trouble to call the police if something seems amiss. The police are grateful, too. It helps to have thousands of 'eyes' around the community.

"Another important part of neighborliness is teaming up with others in block efforts or community projects. It is joining local groups who are fighting for local causes. It is simply *working* to be friendly.

"Everything that brings people together in a neighborhood raises the general level of protection. Alert neighbors, friendly neighbors, responsive neighbors are more helpful in crime prevention than a set of burglar alarms. Cheaper, too. And more fun."

When you are friendly and cooperative with neighbors they will be more inclined to report suspicious happenings at your home when you're away. Through your local civic group or councilman you can help organize a block or community association devoted to your neighborhood's protection. You can pool information about alarms and other safeguards, educate youngsters to the need for vigilance, and, if necessary, patrol streets prone to be victimized. Cagey criminals stay away from such neighborhoods.

A community crime-prevention effort such as "Neighborhood Watch" is administered by the National Sheriff's Association. It goes by various appellations: "Block Watch" in Seattle, Phoenix, and St. Louis; "Neighbor Help Neighbor" in New Orleans; "Home Alert" in Oakland, California.

When neighbors are congenial it's easy to involve them in "Operation Identification," a free antiburglary program in which valuable possessions are engraved with the owner's social security, phone, or driver's license number, kept on file by police. You can borrow the engraving tool at your local police station, bank, or insurance agent (if he's a member of the National Association of Insurance Agents). You'll receive decals to put on your door or window. Your insurance agent will also give you an inventory form to list your engraved objects. If your home is burglarized you'll have a concise record of valuables for the police and the insurance company. As a rule, if you can't prove the stolen property is yours, you can't get it back from the police.

With the decal displayed, a burglar knows that whatever engraved loot he steals from that house will be turned down by fences as too easily traceable. What fences refuse to buy, thieves will not steal. In some 1,500 U.S. communities, Operation Identification has thus proved invaluable to many thousands of homeowners. Surveys in cities such as St. Paul, Minnesota, show that the burglary rate was sharply reduced where the program had been instituted.

Should you have a watchdog?

A dog is probably the oldest burglar-alarm device known to man—and generally the least expensive. Most thieves will steer clear of a house where there's a hound, any hound. The animals are likely to make too much noise, attracting too much attention, no matter how silent the surreptitious entry. For many thieves the risk is too great. A study in one area revealed that less than 15 percent of the homes burglarized had dogs.

With the crime rate rising, hundreds of thousands of jittery homeowners have gone to kennels for trained or untrained dogs. The average family pet can be depended on to bark at strangers. A dog's hearing is far more acute than a human's, and he can identify by smell from yards away. But a dog's value is limited. A smart intruder can carry chunks of delicious meat or other goodies, and the dog may lick the hand that feeds him. More than one clever burglar has rubbed beef bone on his hands and legs to give off a scent that wins friendship. A determined thief can drug a dog, use a fire extinguisher as pacifier, or shoot him. Not long ago, the owner of a Connecticut estate had two Dobermans roaming his fenced property during the night. Both were "neutralized" when a burglar threw sponges soaked in gravy at them. The dogs choked on the sponges.

Liability is another factor. Even the best of trained attack canines may fail to discriminate among a burglar, the mailman, the milkman, and a curious child. Still, a burglar can't know before-hand whether a dog will bite or just bark.

If your home has a large backyard where a dog can exercise, a good attack dog may be a professionally trained German shepherd, intelligent, with a highly sensitive nose. These dogs are used as sentries by the U.S. Air Force. Also consider an Airedale, a giant or standard schnauzer, or one of the larger terriers—provided they have adequate training and a good relationship with you and your family. A Russian wolfhound, Doberman pinscher, or chow can tend to be too aggressive to serve as a guard. An aggressive dog, even with attack training, might injure innocent people—and you'll be subject to a damage suit.

If you prefer a small, barking dog, a Chihuahua is very alert, yelps, and moves fast. A dachshund or a medium-sized terrier will be courageous and emit lots of noise.

Keep your dog inside, especially at night; if possible, give him the run of the house. A large attack dog, held by a chain, can't attack a prowler beyond his reach. Inquire at local kennels for information about professional watchdog trainers.

Remember: When you're away from home for some time, your

dog is unlikely to be around either. And an empty kennel gives a thief the green light.

A perfect watchdog we know is a German shepherd, Prince, whose master, Larry, is a friend of Mac's. On returning to his New Jersey suburban house one evening after a long day in Newark, Larry was concerned when his dog failed to greet him at the door as usual. He called out for Prince. Not even a bark in response. Stepping into his living room, Larry stopped short. A man, a stranger, crouched in a corner clutching his blood-soaked arm; a jimmy lay at his feet.

Standing guard in front of the invader, the German shepherd growled softly, so intent that he did not even turn to acknowledge his master's presence. Larry backed out and phoned the police.

"I used my jimmy on a back window," the burglar told a police sergeant at the station. "Then this dog, this monster—who never made any sound—he grabs my arm. . . ."

At home, Larry patted his faithful watchdog.

"Next time, Prince," he said, "you gotta bite a burglar in the kitchen, where there's a linoleum floor, understand? I don't want the blood of any son-of-a-bitch ruining my living room rug!"

A veteran housebreaker may have an uncanny ability to spot the presence of a dog—so he'll stay away.

The burglar in a Cadillac

One type of elite burglar masks his identity so cleverly that people in the neighborhood are most unlikely to guess what he's up to. Such a superb actor was Howard "The Cat" Turner.

Turner was a short, mild-looking black man in his fifties with one prominent gold tooth that somehow gave him an air of solid respectability. He wore a chauffeur's livery as he drove a big black rented Cadillac around a highly affluent suburb in the early evening. He chose the time when these people would be going out to dinner, perhaps to the theater or a concert. Noticing a departing car, Turner would park his Cadillac near that house, sit at the wheel presumably reading a newspaper but actually watching the house for an hour or so. A uniformed chauffeur was a common sight in such neighborhoods and drew no particular attention.

An accomplished and daring pick man, for years Turner had invaded homes while the occupants were at home and asleep, slipping out with a good haul without awakening his victims. Lately, though, Turner felt he was getting too old to run fast if someone detected him; besides, a police bullet still in his hip had slowed him down. Now he preferred to work in unoccupied houses.

He had a knack for determining when people were not at home. When he was satisfied there would be no interference in a prospective target house, Turner drove his car right up the driveway, stepped out, and rang the doorbell. Receiving no response, he quickly picked the front-door lock and walked in.

Turner knew that people in that suburb, when merely out for dinner or at a party, would probably not bother to hide their valuables; if they did, he knew from long experience where to look.

Sometimes, if before entering he noticed a lighted upstairs window, he would note the name on the mailbox, seek out a public telephone booth, and, obtaining the home's phone number, dial it. If someone did answer he apologized for dialing a "wrong number." More often there would be no answer. Then, he would let the phone ring ten or twenty times, leave the receiver dangling off the hook, hurry to his car, and drive back to the house. If he could hear the phone ringing as he stood outside the door, he'd be sure no one had returned home in the interim. So he'd pick the lock and quickly enter. Inside, he'd take the phone off the hook, wait for a few seconds, and hang up.

Driving around to spot his marks, Turner often used another ploy. He'd keep an eye out for such recreational playthings as boats on trailers or snowmobiles parked in driveways. Then he'd return on Friday night to see if the equipment was still there. If not, he could be pretty certain the owners were away and the house ripe for a visit.

Encountering certain types of alarm systems at a house did not faze this astute burglar. In fact he preferred a house that showed it was "protected" by a keyed shut-off switch outside the front door. Rarely did these switches make use of a pick-resistant cylinder, so Turner would easily apply a pick tool to shut it off. He appreciated these situations because he knew that folks foolish enough to depend for protection on an alarm setup with an outside switch would leave their valuables conveniently laid out for him.

One night Turner spent a little too much time in a house, sensing that he was overlooking a good score. He finally quit at 10:30, which for him and many other burglars is the time to get going—out. In a hurry to leave the scene, he drove down the driveway and into the street too fast, running over a cocker spaniel. Naturally Turner didn't stop but a neighbor witnessed the accident. She took the number on the Cadillac's license plate and called the police to report the hit-and-run driver.

A police radio-car team stopped the Cadillac just as it was approaching the county border. Questioned, Turner blew it: he denied having been anywhere near the scene where the dog was struck. So he was escorted to the police precinct for investigation. The car, of course, had to be checked over and in it police pounced on some jewelry stuffed between the passenger front seat and the backrest. A telephone call to the owner of the home where the Cadillac had come down the driveway subsequently confirmed the burglary. Turner was identified by the witness to the hit-and-run. Arrested, "The Cat" pleaded guilty and was sentenced to prison—again.

Mark at least one object lesson here: Don't rely on an alarm system with a keyed shunt (on-off) switch mounted *outside* your door. An effective alarm should be equipped with a time-delay feature permitting you twenty to thirty seconds after you open your door to get to the control box just *inside* your house. Then you can use your key to deactivate the alarm before it sounds or signals a "break" to a monitoring central office of an alarm company. (For more information on alarm systems, see Chapter 11.)

The Bell System now offers a service called "Call Forwarding," at nominal cost. When you leave home to visit a friend for an evening or weekend, you can preprogram your phone to automatically forward all incoming calls to the place you're visiting. Thus, if a burglar like Turner calls your home number to see if your residence is unoccupied, you can answer—and foil your would-be housebreaker.

Chapter 4 Defending your apartment

STROLLING into the apartment building through the carelessly open service entrance, the bantam, clean-shaven, jaunty young man carried a supermarket shopping bag as he took the elevator to the top floor. Getting out, Tony Mennone moved along the hallway, spotting doors he knew he could easily beat. Under each of these doors—and there were many—he slipped an advertising circular plucked from his bag. A scant corner of each circular was left protruding.

Tony continued the process on lower floors. Before leaving each floor he stopped to look back down the length of the hall. Good, those circulars barely showed; no kid coming down the hall was likely to scoop them up and ruin his plan. Then Tony walked out of the building the way he came in.

That was at 10:00 A.M., long after most tenants had left for work or a shopping trip downtown. At 2:30 Tony returned to repeat his tour. This time he wore a safari jacket with ample pockets. When he came to a door where the tip of a circular was still visible, he consulted a piece of paper on which he had written a list of the building's tenants, taken from the lobby directory. He memorized the name of a tenant on the floor below, in case he was confronted, so he could say he'd made a mistake. Then he rang the bell at the first apartment where his "business card" was still in place, a sign it was probably unoccupied. He pressed an ear to the door, to make sure.

In seconds Tony either 'loided the lock or picked his way in if the door was double-locked. Ten minutes later he was out in the corridor again, one pocket bulging. Within an hour Tony had burgled seven apartments, ambling out with pockets loaded, the half-full shopping bag covered with a newspaper.

Tony Mennone was an apartment specialist working "on the blind" with no specific victim in mind. In high rises, another breed of specialist works only on "fingered" tenants, tipped off about an affluent mark sure to be away from home at a certain time. Such a pro was the legendary Max Pinchek.

Late one morning Max strode into a posh Park Avenue building while the doorman was at the curb summoning a taxi for an impatient tenant. Max, a mild-mannered forty-seven-year-old who looked like the average well-heeled businessman, swung an ordinary man's oversized cardboard suitbox. He was in the building perhaps ten or fifteen minutes. Departing, he smiled and nodded genially to the doorman. The suitbox, no longer swinging, was hanging down in his hand as though heavily laden.

As Max walked to a corner to hail a cab, two detectives (Dick Weber and Tom Sullivan) who had been tailing him stepped to his side. With his free hand Pinchek reached under his coat, then stiffened when one of the cops ordered him to take his hand out "very slowly." There was a gun inside Max's waistband.

At the police station the suitbox was found to contain a woman's sable fur coat. His pockets were crammed with jewelry worth some $18,000, two picklocks, a tension bar, and a shoe horn. Why the shoe horn? It makes an excellent small jimmy, handy for opening locked jewelry boxes. The gun, rare among so-called gentlemen burglars like Pinchek, indicated he could have been dangerous.

The apartment he had chosen was tenanted by a Mrs. Walker, the estranged wife of a Wall Street broker. She had been fingered by Max's favorite criminal receiver, Al "the Angel," who ran a small jewelry shop. Al, who once sold Mrs. Walker a diamond pendant, had read a gossip column item about Mrs. Walker attending the Cannes Film Festival.

As for the $18,000 in stolen jewelry, Mrs. Walker later told police she had not taken the pieces along on her trip because they had been given to her by her mother and were not in "stylish settings." Before she left on the cruise Mrs. Walker had concealed her jewelry box in a bathroom hamper, covered with soiled linen. (Max, of course, was familiar with that ploy.)

Security-minded though not too knowledgeable, Mrs. Walker had had a handyman install another lock on her door. It was a good dead-bolt type, the cylinder protected from forcible removal by a cylinder guard ring. It would have been difficult to jimmy it or pull the cylinder with vise-grip pliers. But scratch marks detected on the cylinder pins proved that Pinchek had used lock-picking tools. Confronted with the evidence, Pinchek boasted the job had taken him less than a minute.

In sharp contrast to door shakers and the young man who used the circulars ploy, Max was among an elite corps of burglars. For at least fifteen years he had managed to operate as a deluxe pick man without incurring a police record anywhere. A most unlikely thief, he owned a small chain of dry-cleaning establishments, probably purchased with the proceeds of his rip-offs, and lived with his wife and two children in an upper-middle-class neighborhood in New York.

Four times a year Max took two weeks off to fly to Dallas, Los Angeles, Seattle, and other cities where he performed his jobs in high-rise apartments and at hotels. He hit even Las Vegas, where security is particularly good and the consequences dire if caught. In Chicago he would stop at a locksmith supply house to replenish his kit of hand picks, tension bars, and other tools of his trade. After his out-of-town forays he immediately deposited his scores with his fence, Al the Angel. It was through an informer who also traded with the Angel that detectives latched onto Max.

Pinchek's possession of a gun had an interesting background. Four years before, just about the time he started traveling around the country, Max had attempted to burglarize the tenement flat of an ironworker reputed to have a lot of cash at home. When the tenant and his wife walked in and surprised him at work, Max proceeded to shoot the man three times in the stomach, and scrammed. Rushing down the rickety stairway he fell, dropped the gun, and ran off without it.

The gun, a .32 (the same caliber as the gun found on Max by Detectives Weber and Sullivan after the Mrs. Walker burglary) was processed at the time for latent prints, and a partial print was lifted from the barrel.

When Pinchek was fingerprinted, Detectives Weber and Sullivan recalled the shooting with a .32 four years before. The police lab reported that the partial print on the old weapon matched the print of Pinchek's right index finger. He was put in a lineup. The ironworker, who had miraculously survived his wounds, was brought to police headquarters, accompanied by his wife, in an effort to obtain an identification.

Yes, they said, this man very much resembled the burglar they saw so briefly that night—but they couldn't positively say he was the assailant. Under the law, to be admissable as evidence fingerprints must meet certain criteria, a given number of points of identification that may be gleaned from a fingerprint. What the police had was a "partial" on the gun used against the ironworker. And that partial print did not contain the required number of points. Thus Pinchek could never be charged with the shooting.

When you put a Max Pinchek under a magnifying glass and study what makes him steal you'll find that whatever motivates him doesn't differ much from the emotional factors impelling a stickup man. Both criminals have about the same degree of contempt for the rights of others and both are basically hostile. While the stickup man openly grabs at his loot, ripping it out of your hands, the burglar furtively slips it into his pocket when your back it turned.

Pinchek had grown up in a normal household of hard-working immigrant parents and felt no more economic pressure than the next fellow; it wasn't hunger that pushed him into the first apartment he invaded. Instead, this man, with better than average intelligence, had made a conscious decision to steal—and to steal by stealth, sharpening his skills to lessen the chances of being caught at it. He further insulated himself by dealing with only one carefully cultivated fence and avoiding the company of other burglars. And he carried a gun, the final tool to prevent capture and his picture being splashed in the newspapers.

But what does Max do when he realizes he can't extricate himself, that he is really caught? He boasts of how quickly he got into the Walker apartment. This, however, was only a momentary slip, an aberration.

Pinchek's masks stood him in good stead when he appeared before the judge for sentencing after having entered a guilty plea. Despite the probation report prepared for the court, detailing everything the police discovered about this busy thief, Max

Pinchek—a "responsible businessman" in his community—was awarded a suspended sentence!

The urban burglar

Too often, city people living in a large multiple-dwelling complex guarded by a doorman have a false sense of security. Smart burglars get by the doorman by presenting a smooth front or, like Max Pinchek, waiting until he is diverted. Service and garage entrances may be left open or their door locks can be readily defeated. To urban thieves such high rises offer many advantages. Generally, most of the tenants are unknown to each other, so a new face in the lobby elevator or hallways seldom arouses suspicion.

Apartment rip-off men, like Tony Mennone, tend to make their bold moves weekdays between 1:00 and 3:00 P.M. Burglary statistics bear this out. Almost universally a homemaker does her housecleaning chores in the morning; after lunch she goes out, to shop, visit friends, carry out errands. True, in many apartments both the husband and wife are away at work all day but burglars on the blind don't know which these apartments are. Those who dare to function evenings most often are working with specific information about marks.

Billy K. rents a grocery delivery bike, packs it with half-empty cartons of groceries, and wears the kind of clothes worn by neighborhood delivery men. So he arouses no suspicion in a doorman when he enters through the service door. For a change, he will pose as a repairman, have a pencil tucked behind an ear, hold a clipboard, and lug a tool kit. Words such as "Ajax Repair" on the back of his shirt are the only credentials he needs.

When a top-security building maintains TV cameras for surveillance in hallways, a canny burglar will have a confederate at his rear (or side) so that his fast job on a door will be blocked from the camera's view. If he is a crude thief using a jimmy, he won't risk this, but a pick man will.

Apartment thieves prefer to work high-rise elevator buildings because the scores are generally better. They like to be near fire stairways so that they can duck there if they hear an elevator approaching the floor.

When a pick man hits an apartment building he usually does a half-dozen burglaries the same day. Victimized tenants, believing that keys were employed, blame management.

In an apartment house without a doorman, many burglars either use a celluloid or merely press several buttons downstairs at random until someone releases the front-door lock, enabling them to enter and wander through the hallways for a potential victim. If all apartment doors have peepholes, burglars sometimes tape the lenses on other doors while breaking in, then remove the tapes later. And there are the multiple other ruses and stratagems we have mentioned previously.

PRECAUTIONS TO TAKE

Here are some ways that tenants can discourage or repel would-be burglars:

► Even if you are just going down the hall to a neighbor "for a minute," always lock your door. You may be gone longer than you expect to be.

► Avoid leaving your keys in a pocket of your coat that you may place in a checkroom or hang in a public place.

► Teach your children to be careful with all keys.

► Never label your house keys with your name and address. If you have to tag them for any reason, use your business address.

► On moving to a new apartment, if possible change the cylinder or add a deadlocking auxiliary lock. If you are required to give the building superintendent a duplicate key (for emergencies) place it in an envelope, initial it across the flap but don't write your apartment number, and seal it with Scotch tape. If there's a break-in and the envelope has been tampered with, the police will be interested. Your super should keep tenants' keys in a locked box, coded and not identified by apartment number in case someone gets at the box.

► If you don't already have a door peephole, install one. Make sure it has a wide-angle 180° lens that will reveal anyone who might be hiding outside your normal range of vision.

► Going out evenings, leave a light on where it shows under the door in the exterior hallway. Leave a radio playing so that it can be heard by anyone listening at your door.

► Do you lack control over the locks you can use on your exterior door? Many cities now allow your own lock despite adverse terms in a lease. You can make your bedroom intruder-proof. Install a good inner lock and/or sliding barrel bolt. Leave the key in the lock in the event of fire, and leave a duplicate with a trusted neighbor in case you're taken ill.

► Have a telephone extension in your bedroom. Use it if you hear an intruder in your living room.

► Install a strong short chain guard that permits you to open the door (after preliminary screening with your peephole), but not so wide that an invader can reach through the opening to grab you.

► Select one good burglar-resistant lock for your front door. (For more information, see Chapter 11.) Burglary *can* be stopped and the easiest to stop is the pick man.

► When you use double-cylinder locks on your door, mounted with one-way screws, if a burglar enters through a window he won't be able to leave except by the same window and can't carry out as much loot as he could through a door.

► If any of your windows can be reached by a thief—particularly a window facing a fire escape or near a ledge or abutment—install a non-key-operated folding or accordion-type gate approved by your fire department. Make sure each member of your family knows how to use the escape feature if fire breaks out. Permanently installed bars across a window constitute a hazardous trap in case of fire, and most cities outlaw such immovable window bars. State law may allow window locks as long as they don't block the "egress" window.

► For a woman living alone: Use only an initial for your first name on your mailbox, doorplate, the apartment registry listing, and the telephone directory. Have a phone in your bedroom, and a lock on your bedroom door.

► Get to know your immediate neighbors. Agree to phone each other if either of you hears suspicious sounds. If you (or your neighbor) don't answer the phone, the police should be called. Attach the police number to your phone; in most communities it's 911.

► Will you be away for a long period? Consider subletting your apartment.

► Especially if your neighborhood has been plagued by burglaries, get involved in a community crime-prevention effort or program such as "Block Watch" and "Operation Identification" (see page 73, Chapter 3).

What about a watchdog in your apartment? A large dog, such as a Doberman or a German shepherd, rarely adjusts to close confinement. Women living alone may feel safe with a fierce dog like a Great Dane or a boxer, but not all are effective city watchdogs; some don't bark but wait for an intruder to enter and then pounce. If you feel you should have a canine-alarm, a dachshund has a loud bark, and fox terriers and Airedales are usually brave and alert.

In a Philadelphia apartment house not long ago a toy fox terrier began yapping at two in the afternoon and wouldn't stop. A neighbor phoned police to complain of the noise. The cops arrived and nabbed a burglar leaving the building. He was loaded down with loot from four apartments. Although he hadn't tried to break into the terrier's apartment, because of the barking, the thief's activities in other adjoining apartments was enough to trigger the canine-alarm mechanism.

Among the strange specialists among thieves is the dog-napper, the burglar who steals expensive pedigreed dogs. Gabe Kunja, a diminutive, craggy-faced character known as The Dwarf, had been a trainer of show dogs, working for several prominent families. Probably bored with, or weary of, that occupation he turned to larceny and was twice arrested for housebreaking.

Came a time when police in Manhattan received a number of complaints about pedigreed dogs missing from locked apartments and town houses. It developed that Kunja would walk along Park Avenue, spot a potential four-legged target on a leash with its master or mistress, and follow them home. Later he used a pick to break into the residence and lift the dog.

When Kunja was finally tracked down, police found in his apartment stacks of forms and two embossed seals from kennel clubs. Using bogus papers, Kunja gave his canine loot new "bona fide" identities and sold them at fancy prices. Also uncovered in his apartment were three small leather cases containing picklocks, as well as assorted leashes and dog yummies.

A parakeet—yes, a parakeet—once turned out to be an effective alarm system. When Mrs. Helen O'Reilly came home from shopping one afternoon she found her apartment door locked and Fred, her small green and blue parakeet, "screaming his head off." Fearing to enter, she ran to a neighbor and called the police. They put out a radio alarm and caught two young burglars ransacking and vandalizing the flat—but Fred was dead, his "alarm" turned off. A few days later employees of a neighborhood pet shop presented Mrs. O'Reilly with a new parakeet. "That bird, Fred," said one of the shop's clerks, "saved her life. We know how sad she must feel so we thought she should have another bird." Mrs. O'Reilly is proud of her "watchbird."

A good watchdog, of course, will deter many thieves. There's always the fear of being attacked and even the unreasonable anxiety about needing rabies shots. Two of the boldest and busiest specialists in apartments and hotel rip-offs, the team of Al Perenzo and Stan Lipman, committed dozens of burglaries but never where a dog was involved. Robert McDermott, who played a key role in finally busting this Odd Couple, relates their saga:

The versatile losers

Although it is certainly true that a skilled pick man like the durable Max Pinchek does not need any help on the job, few professional practitioners work solo. No matter how experienced the man may be, while he's operating on the lock—and after he's inside—that jumping pulse in his throat is damn near choking him: He is scared.

So the burglar wants company. It's nice to have someone's hand to hold, figuratively, when you hear a sudden sound out in the hall and you breathe, "What was that?"

Partners in burglary spend considerable time together and when apart they are constantly reaching out to each other by telephone. I found it so with such pairs as Bernard L. Foley and Cal Davis (forcible entries in private homes), David Berg and Peter Giovetti (hotels, apartments), and many others of their tribe.

Working in tandem these thieves tend to depend on and trust each other. But one peculiar partnership I encountered was based on fear rather than mutual faith. It was the unlikely team of Stan Lipman and Al Perenzo.

Lipman, then thirty—soft-eyed, curly-haired, with a kind of lopsided face—had been a private insurance investigator duly licensed by the State of New York. An electronics wizard and expert wire man—planting microphones and tapping phones—he was also hired in divorce cases, for industrial spying, and the like. With the celebrated Steve Broady, he had been enmeshed in the criminal bugging of New York's City Hall during Mayor Robert Wagner's administration.

Lipman was a half-assed burglar, in my judgment, an insufferable braggart always hell-bent for the big score.

In sharp contrast, Perenzo, a husky muscle man with a stony look about him, struck me as a complex bag. A graduate of New York University, an erstwhile professional boxer, he held a Purple Heart and Silver Star for World War II service with the 82nd Airborne in Europe. He was said to be the secret owner of C'Est Si Bon, an elegant East Side supper club investigated in a case of bribery to fix a liquor license. Years before, after an assault and robbery arrest, he had languished in a penitentiary for twenty months.

A flashy Broadway type, Perenzo, thirty-seven at the time, wore tailor-made silk suits and custom-made shirts, liked to exhibit a fat roll of currency, squired a succession of glittery show girls. Surly and close-mouthed, he could be charming in an asp-like way. Cops and crooks alike knew him as a Turk, an uncontrollable wild man, vicious and sadistic. Few in the burglary business would go near him. Except Stan Lipman. Evidently both needed a handholder on jobs.

Unlike run-of-the-mill burglars, our Odd Couple followed no fixed modus operandi. Where lock cylinders were vulnerable they resorted to 'loiding, used master keys or duplicates of tenants' keys. Among other tricks of their trade, on occasion they would drill-and-shim a victim's cylinder. They carried "manipulating keys," thinly filed standard keys that would work if a cylinder had been extensively mastered, as in apartment buildings or large hotels.

Since the team was not especially adroit at any of these techniques, occasionally they shifted to robbery. For a direct approach

Perenzo packed a .22-caliber Colt Woodsman. In a residence robbery they rang a doorbell. When the occupant answered, Perenzo pulled his gun and they stuck up the place. People were handcuffed or trussed up with wire.

After Lipman did a job or two with Perenzo he found himself locked in. It started with Stan supplying information on marks he had had in mind for some time. At first wise-guy Lipman liked the notion of being hooked up with the tough Perenzo. Then, gradually, the Turk began calling all the shots, pressing for more and better scores, now and then revealing the unmanageable, violent nature that had made him an underworld outcast.

Lipman, though brazen enough for bugging and burglary, soon realized he was no match for his partner. He now had to worry not only about an arrest but also about the consequences of Perenzo carrying a gun "just in case." Lipman began to fear that Perenzo would shoot *him* for some crazy reason. Still, Stan was earning a good living at stealing and enjoying the fruits of their alliance.

The Lipman-Perenzo sprees broke out at a time in 1963 when the juicy publicity about burglars' big coups attracted swarms of thieves—pros and amateurs—to the Big Apple. The media kept referring to burglars as "phantoms," "Raffles," and latter-day Jimmy Valentines, ascribing to them talents that allowed them to walk through walls and float through ceilings. That year an estimated $12 million in jewelry alone was reported to have been stolen from individuals only in New York City.

It was the time when the Boston Police Department sent me lock cylinders to determine whether the Boston Strangler was a pick man. Homicide men there could not believe that all the murdered women would admit a stranger. I reported that none of the cylinders had been tampered with.

It was a time of burglary hysteria when I was interviewed repeatedly by reporters and magazine writers, even by the *Wall Street Journal* and Johnny Carson, advising citizens how to protect themselves against intruders. About that time, the 563-carat Star of India was snatched from the Museum of Natural History by Jack "Murph the Surf" Murphy, who simply climbed through an unlocked, unalarmed window.

Other than this case, much of the epidemic we gradually attributed to Lipman-Perenzo. Although they also hit hotel suites, mostly they focused on high-class apartments. These were some of the burglaries and stickups the bold and busy pair was suspected of having perpetrated:

> Zsa Zsa Gabor, at the apartment she shared with her current husband, Herbert Hutner: five oil paintings (including her portrait), valued at $10,000.
>
> Actress Janet Leigh: $9,000 in jewels from her Regency suite.
>
> Mary Livingston, wife of Jack Benny, robbed of "more than $200,000" in jewelry at the Pierre.
>
> Mr. and Mrs. Frank Loran (he was a stockbroker) were celebrating their wedding anniversary at their luxurious co-op apartment. Two gunmen, posing as deliverymen, forced an elevator operator to take them to the apartment, then shoved him in when the maid answered the doorbell. The robbers, holding five persons at gunpoint, scooped up $55,000 worth of jewelry, ran downstairs, and hailed a cab.

The case of Mrs. Sofie Merrett, wife of a prominent businessman, generated more than routine drama. One afternoon two detectives in a squad car heard on their police radio: "A baby on an apartment building ledge!" At the scene, a building employee in the backyard had his arms outstretched to catch a bawling child from a seventh-floor open window. From the window, screams split the air.

When the detectives rushed to the apartment they found two hysterical women, Mrs. Sofie Merrett and her mother, handcuffed to a radiator. The eighteen-month-old child was grabbed from the window.

The women related what happened. Two men who claimed to have been sent by the telephone company had entered, one of them producing "a long black gun." Masking themselves, the men scoured the bedroom, picking up $16,000 in jewelry. One of the stickup men demanded a solitaire Mrs. Merrett was wearing. When she balked he threatened to break her finger. His partner, impatient, kept urging, "Out . . . let's get out . . . leave it, we have enough!" (Lipman?) As they left, the tough guy (Perenzo?)

swore he'd return and shoot if they cried out before the men had time to make their getaway.

When I was assigned to this case I found the lower pins of the door's lock cylinder deeply scarred by a heavy-handed pick man. Apparently those fellows had first tried burglary but the lock had beaten them.

In another solitaire caper the duo took a quite different tack. Laura Brand, the proprietor of the Latin Quarter, a popular nightclub, sported a twenty-five-carat solitaire that caught Lipman's greedy eye. He had been introduced to Brand by a hustler who frequented the club. Stan eagerly reported his solitaire target to Perenzo, who promptly approved the enterprise.

Stan the Lothario, dating Laura, took her out for a night on the town one evening. The Turk used the occasion to beat the lock on her suite at a residential hotel. Despite a thorough search he came up empty-handed except for a few trinkets. She had worn the ring on her date.

Hours later, when the team got together, Perenzo insisted:

"We're going back right now. That hotel lock is a cinch and we know she's got that stone."

Jittery, Stan protested. "Hey, she knows me—how in hell do we get around that?"

Perenzo wouldn't listen. "She doesn't know I was in there. You stay away from her—I'll tackle it and we'll use ski masks and the rugs [wigs]."

Back they went to the hotel, picked the lock, donned masks, tiptoed into Brand's bedroom, and scooped up her solitaire and other jewelry for a total of $90,000.

(Later, Laura Brand told us she'd been awakened by sounds in her room and could hear two men whispering. Her eyes covered by a sleep mask, she lay terrified and mute, pretending slumber until the outer door closed.)

Next day, I automatically made a microscopic examination of the lock cylinders. Sure enough, the lock had been picked. I did not need a microscope; the followers (lower pins) were so badly scratched that the pick traces could be seen with a naked eye. It must have taken Perenzo at least ten minutes each time to get that "cinch" of a lock to go for him. An expert burglar the Turk was not.

For four months, while all these and sundry other break-ins and stickups had been going on, detectives in the East Side's Seventeenth and Nineteenth Precincts were catching hell. So many jobs were being reported that detectives could not concentrate on any one of them. They felt as if all that could be done was to keep score.

Inevitably we in the Safe and Loft Squad became aware of Lipman and Perenzo. Even before the Laura Brand take, some of our best tail detectives had tried surveillance. But the pair proved too slippery. Sometimes they'd spend a day or two doing nothing but watch out for tails.

Lipman, particularly, behaved like a hen on an egg. Just walking his laundry to the cleaners he constantly looked over his shoulder. Together, the pair would take a cab, drop it for a subway, let trains pass by, then step into a car just before the doors closed. So our detectives had to quit them periodically, wishing they would cool off.

Of course, I "visited" both Lipman's and Perenzo's apartments, both in the East Sixties, while they were out, hoping to find their loot. No luck. Not even burglars' tools. The places were clean, except for some men's wigs. While both men had high foreheads, neither needed a toupee; so the presence of the "rugs" was suspicious.

It was a downright frustrating—and embarrassing—situation for all of us in the Police Department. Until the day the Odd Couple finally tripped up and botched a job.

One morning, plucky sixty-two-year-old Mrs. Morris Lipschitz, wife of a telephone-company executive, was in her Central Park West apartment when a phone message informed her that two company employees would come to install a new white phone in a half hour. When the men arrived, their bizarre behavior and language—and the way they immediately ripped out her phone wires—aroused her apprehension. Slipping out of her apartment, Mrs. Lipschitz told her next-door neighbor she was "nervous"—and to "listen."

As soon as Mrs. Lipschitz returned to her apartment, the men announced: "This is a holdup!" and flashed a gun. The lady fairly shouted, "Don't hurt me, I'll give you what you want!"

Forced into a bedroom, she handed over sixty-four pieces of jewelry.

Her neighbor, listening, alerted the doorman and called the police.

In the street, cops converged on the two men as they ran in different directions. Perenzo, heading west, carried a black bag, later found empty. He was so panicked that his loaded .22 automatic was still in his other hand. A patrolman caught him without a struggle. Stan, running south, was seen by a passing motorist who kept blowing his horn as he followed the man. Though Stan hopped into a cab, he was soon captured a few blocks away. The jewels were recovered in a hallway where Stan had stashed them.

But where had all the other swag—the multiple scores from Zsa Zsa Gabor, Janet Leigh, and the rest—been hidden?

A month or two before the arrest, Detective Joe Kiernan had called me to say, "We need your hands." He had caught a homicide case and wanted me to open the victim's apartment to see what could be turned up. Before I accepted any such job, I always subjected it to a hard-and-fast rule: I must have some safeguard against being used for something not strictly kosher. So any detective drafting my services had to carry the blessing of his squad commander or someone higher up. After all, the Detective Division is made up of people with human frailties, and every once in a while we turn out a rogue, our own Turk. In this instance, of course, Joe Kiernan had the blessing I needed.

After I entered and left the murdered man's apartment, defeating a hundred pounds of locks (ranging from Segal drop bolts to "police" locks), Kiernan and I went out for coffee.

Gradually we got around to exchanging notes and theories on the Odd Couple. I knew that Lipman was beginning to come apart a little at the seams. Besides drinking too much, once in a while he partook of some of the Turk's private stock of cocaine. His joining up with Perenzo had developed a serious unforeseen drawback: receivers refused to handle their swag. So they were accumulating a fortune in stones and other valuables, rather than dollars. Perenzo, unstable, would take a big fall someday and anyone convicted with him would also go down hard.

That day with Detective Joe Kiernan, we agreed that Lipman and Perenzo had probably executed the Laura Brand and Zsa Zsa capers.

"Mac, the thing is," said Kiernan, "this crazy bastard Perenzo is going to shoot soon. We have to get them. We may even wind up with a dead cop."

"What is Laura saying now about the voices she heard during that night. Did she recognize Stan's voice when the guys were in her room?"

"She says she didn't. But I know this: Lipman called her, said he'd read about her burglary and asked for another date. She said no. By the way, Lipman is putting the word around that he and Perenzo have a cache of the stuff they've got and he's rigged up the lock so that if it's picked a shotgun blast goes off. Maybe other guys would like to find that stuff; he's afraid of a hijack."

"A stash with *a* lock rigged up or *the* lock? Because if he said *the* lock it's probably an apartment."

"Yeah, it's an apartment. As I heard it, the stash is in another name, so it must be an apartment."

"I'm not sure it's a bullshit story for thieves. Joe, they know what I do, using burglars' own tricks against them. I can see how the booby trap could work and I'm not into electronics. Maybe with a microswitch. So Lipman says the shotgun goes off if the lock is picked? That means if a *key* is used, the thing doesn't go off, right?"

"If it's an apartment I guess so, or Lipman and Perenzo themselves couldn't go in. But, Mac, to me it looks like if you use a pick it's the same as using a key."

"It *accomplishes* what the key does but not in the same way. Maybe I can handle it. After I get all the pins up I won't turn the keyway at all until I block the bottom of the keyway with the back of the pick and press the pick in like it's a key. But Joe, I'll be scared spitless. I'd like to stand Lipman in front of that door when we find it and pick the lock under his arm."

"This story must be just for the street guys, though. If we find the place, he's finished anyway. Especially if that shotgun goes off on you, Mac, then he'll be dead, too."

"Thanks. So you're going to avenge me."

"You'll wear a bulletproof vest and the iron hat."

"I did that once in a Brooklyn flat. In the Willie Sutton case I opened a door that Arnold Schuster's killer, Freddie Tenuto, was supposed to be behind with a gun. The damn bulletproof vest weighed about a ton and it took me twenty minutes to pick an ordinary Yale. We're going to get Lipman and Perenzo. I just hope it's only Perenzo who bleeds."

We left the coffee shop in a downpour. I remember shivering in the warm rain.

The day after Lipman and Perenzo were collared in the Lipschitz stickup fiasco, I was in Summit, New Jersey, lecturing at the Bell Laboratories on "Surreptitious Entry and Safeguarding Secrets." A phone call from Lieutenant Stephen McGrath of the midtown Seventeenth Precinct summoned me back to Manhattan. The hot news: Two detectives, Raymond Devine and James Biddescombe, were pretty sure they had pinned down the Odd Couple's cache apartment on East Sixty-fourth Street. A "confidential informant" had provided the lead.

Photos of Lipman and Perenzo had been shown to the building's doormen and tenants, who identified the target apartment on the fourteenth floor.

I showed up at 5:00 P.M. Awaiting me was Lieutenant McGrath, later joined by a police captain, Joe Kiernan, and three other detectives.

The lieutenant told me that Perenzo had rented the place a couple of months before under a phony name, but that we couldn't move until we got a court order authorizing us to break in and search. Devine and Biddescombe were at that moment trying to find a judge to sign the order. It had taken all day.

McGrath told me there were three locks, besides the house lock, and no alarm device. He would not call Emergency Service to ram the door down. If nothing incriminating was found in that apartment there could be trouble with the judge who issued the search order.

I pointed out that if a ram were used and Perenzo had the trap rigged as a normally closed circuit, tearing the door open would almost certainly break the circuit and it would go off.

"I think I can handle it," I said.

Three hours later the court order came through. We all took an elevator up. At first glance the locks did not look like anything extraordinary.

The apartment door was situated in a corner. With a wall on the right I could get to the door only from the left. Maybe, I kept thinking, Lipman knew I'd heard of his trap. And knew I would stand to one side, the only side, and the blast would come through that wall.

The other six cops stood well back behind me, out of range of any possible blast. Under my breath I prayed to Saint Dismas, the patron saint of thieves: "Good Thief, this is a good work. Get me in quickly—and safe."

My arms fully extended, I picked the three dead bolts, one by one, that had been installed to beef up the house lock. In each extra lock I simulated the presence of a key by pressing the pick to the bottom and back of the keyway. I kept pressure on the plug and "rode" my pick and tension bar in a 360° arc. As each of the bolts dropped I realized what was meant by "dead silence." Finally I used a simple celluloid strip on the house lock.

Nothing happened. No shotgun blast. Evidently Lipman and Perenzo were no more clever at figuring out a trap than they were in keeping me out.

Cautiously we stepped into the Odd Couple's quiet hideaway, what a front-page *Daily News* headline next day would call a "Robbers' Roost." Facing us on a wall in the living room was a large framed oil portrait of Zsa Zsa Gabor displaying her pronounced cleavage. Lush rugs, far-out furniture. An ultracreative interior decorator must have had a gorgeous time fixing up the place.

After a quick look around, we focused on a closet with a lock the pair apparently considered foolproof.

"Damn it," I mumbled to McGrath. "Maybe this is the trap."

Again I silently appealed to Saint Dismas. I had no trouble opening the closet. Inside we found this inventory:

> About one hundred pieces of jewelry—including rings, settings, bracelets, earrings—later valued at $110,000. Most of the baubles were to be identified as the take from the Loran and other recent burglaries and holdups.

More than one hundred tagged keys from various hotel rooms and private apartments.

The cylinder removed from the door lock of Zsa Zsa Gabor's apartment. Lipman and Perenzo had taken impressions of it and were in the process of hand-cutting a master key for all apartments in her building. Four other Gabor paintings.

A violin-shaped money clip inscribed, "To Jack Benny from the Friars."

An assortment of wigs.

Detailed layouts of many posh East Side apartment houses.

A sizable quantity of narcotics and drug equipment; porno films and photographs.

A wide array of delicate, expensive burglars' tools.

Papers indicating Perenzo maintained safe-deposit boxes at two banks (later found bare).

Cuff links bearing the name of Leon Levy, wealthy Philadelphian and president of the Atlantic City Racing Association. (A curious trophy, as we shall see.)

It seemed clear to me that our Odd Couple did not know what to do with all their swag—or even with whatever cash proceeds they managed to harvest. From a bureau drawer we dug out five hundred pari-mutuel racetrack tickets from Aqueduct, Roosevelt, Monmouth, and other tracks. The total purchase value, at least $75,000. Every ticket was on a loser!

They were losers, all right. Earlier in the day we crashed the apartment, two strangers had been seen by tenants hurrying from an upper floor of the building. One man carried a suitcase, the other a shoebox, and they left in a car bearing New Jersey license plates.

When word of this development leaked out, newspapers estimated the suitcase/shoebox contents at $500,000—a guess based on the Lipman-Perenzo links to recent larcenies. Despite persistent police investigations, the strangers were never located. One theory was that they had hijacked the Robbers' Roost, cleaning it out except for the locked closet they could not penetrate and the far-out furnishings.

Nabbing Lipman and Perenzo solved a total of thirty-four crimes. At their trial in State Supreme Court, Lipman was slated to testify against his partner. The Turk had been out on $61,000 bail. On the first day of the trial a sensation erupted when Assistant District Attorney Norman Ostrow revealed that he had tapped the phones of the prosecution's chief witnesses.

In a scene reminiscent of a television drama, Ostrow told the court:

"Plans are under way at Perenzo's urging, request, and insistence that certain funds be obtained so that the prime witness (Lipman) may be disposed of, either seriously injured or killed. I have two sources to this and I am ready to produce them in court if need be. There is no question that Lipman's life is in danger. Perenzo is ready, willing, and able to destroy Lipman or anyone else who gets in his way."

On one occasion, Ostrow added, a Molotov cocktail was tossed into the lobby of a building owned by Mrs. Lipschitz's husband. A scheduled witness against Perenzo, she was distraught enough to disappear for three days.

After listening to Ostrow's tapes of the taps, Judge Arthur Markevich was shocked. "This makes me physically sick," he said. Revoking Perenzo's bail, the judge adjourned the case for a week.

When the trial resumed, the Turk acted up, yelling that he was being "railroaded," and otherwise disrupting the proceedings. After repeating warnings, the judge ordered Perenzo gagged and bound in a straitjacket.

Convicted, Al Perenzo was hit with a fifty-to-seventy-year stretch for the "protection of the community." Lipman, pleading guilty to one of the burglaries, and having cooperated with the prosecution, got off gently with a suspended sentence. The Odd Couple had finally been awarded a legal divorce.

Saint Dismas has always stood me in good stead in my war on burglars. I wear his medal not only to foster success on my job but also because I know that burglars *are* vicious. Only once was I hurt and then not seriously. I've been lucky.

Since burglars tend to be dangerous, you need more than the hardware. You should cultivate the good habits we recommend for your security. Invariably when the Lipman-Perenzo team—frustrated by good locks—could not score with burglary, they resorted to one or more subterfuges. They didn't always succeed. Some of their intended victims, using the chain bolt on their doors, thwarted them by keeping them standing out in the hallway. "I'm going to call your company," the tenants said, "to check you out. You don't mind, I'm sure." On these occasions the Odd Couple turned away from the doorway and walked off.

Chapter 5 When you're away on a trip

TO CELEBRATE their silver wedding anniversary, Mr. and Mrs. Cloyd Hazelton of Racine, Wisconsin, decided on a trip to Europe. Stopping over in New York for a week of shopping and theatergoing, they checked in at a renowned hotel. On their second day they visited Cartier's where Mr. Hazelton bought his spouse a $15,000 diamond ring, a seven-and-a-half-carat brilliant cut solitaire.

That evening in the hotel restaurant, Mrs. Hazelton kept flashing her ring pridefully at their table. Going out to the theater, the couple felt it wouldn't be prudent to have the lady wear her new ring on the streets of crime-infested Times Square. So she put her trinkets in a small jewelry case on the dresser. She hid the new solitaire, cleverly she thought, in the pocket of a robe hanging casually in a closet near the door. The couple carefully locked their door and left to enjoy the amenities of the Big Apple.

On their return from the theater they were flabbergasted to discover that the jewelry case was no longer on the dresser. They found it, empty, on the closet floor. And the pocket of Mrs. Hazelton's robe was also empty, the $15,000 ring gone.

A somewhat similar burglary occurred some months later at a New Jersey hotel. A newspaper photo had shown wealthy Beatrice E. Fleming wearing what the caption referred to as a fabulous Indian necklace. The caption also revealed the hotel where she was staying. In her absence one afternoon, two burglars slipped into Mrs. Fleming's hotel suite and walked out, their pockets swollen with more than a million dollars worth of jewels—and a piece of celluloid.

Mrs. Fleming, unlike Mrs. Hazelton, had merely slammed her

door, without using her key, when she left. Both had neglected to place their valuables in the hotel vault. Across the street from Mrs. Fleming's hotel a bank displayed a familiar large poster. It depicted a burglar climbing out of a window with a bulging pillowcase over his shoulder. The poster's legend read: "Why Take Chances? Use Our Safe Deposit Boxes."

Ironically, Mrs. Fleming served as board chairman of that bank. Informed of the burglary, her deepest concern was for a ring worth less than fifty dollars that held special sentimental value for her.

If Mrs. Fleming's burglars had not been caught in the hotel lobby—and if the lock on Mrs. Hazelton's door had not been examined to reveal the minute pick marks in it—both of the break-ins would probably have convinced the ladies that hotel employees were responsible for the thefts.

Away from home, on vacation, for business, or otherwise, bear in mind the admonition of veteran burglar Danny "Sticks" ("sticks" is underworld jargon for lock picker):

"The doors at many hotels are a joke. You could blow some of those doors open with a single breath."

Doors in too many hotels and motels depend on an easy-to-'loid spring latch, locks that are mastered and simple to pick—and the hostelries are using the same keys that have been "lost" over and over again. Certainly one of the easiest ways to burglarize a hotel room is to use a key. Since burglars know that many hotels don't bother to change the lock if a key is not returned, any key for a hotel with the room number stamped on it is prized. Prostitutes are an excellent source of hotel keys and they're paid for them by burglars. A missing maid's key, a housekeeper's key allowing entry into all rooms on one or more floors, or the hotel's Grand Master Key that is missing—any of these will surely not cause the hotel to rekey the entire building.

Recall poodle-walking Karen Dosche's frustration over the fact that some of the tenants in her building had installed auxiliary locks, making her "master key" for their house locks useless. Her lock-wise accomplice was unable to help Karen where her neighbors had installed their own locks; otherwise she and Shorty would have had even more of a field day.

In hotels the locks on interconnecting doors inside the rooms are on the same master key system as those throughout the building. So a burglar simply has to pick the lock on one of those inside doors, take the cylinder apart, and by a process of elimination make a key for the entire place.

All kinds of hotel staff people—electricians, maids, and handymen as well as desk clerks—have room keys. Some authorities estimate that four out of five thefts from hotel rooms are by dishonest employees. We don't agree. Most employees are carefully checked out before they're hired and relatively few are dismissed or arrested for theft. While it is true that hotels are reluctant to publicize any misbehavior of employees, in very few cases of the hundreds McDermott investigated were hotel workers found to have been involved.

Hotel security men? Most are retired police officers or FBI men, their principal duty to maintain the hotel's reputation by handling too-boisterous guests and unwelcome hookers.

Occasionally, by one means or another a hotel worker without law enforcement background will be appointed to the security staff. Later, after he has rubbed elbows with city detectives for years, it will be forgotten that he was untrained. Then the day may come when a really heavy burglary occurs. The door lock is examined to determine how a thief entered and that "security man" says to the investigating detective hunched over his microscope: "What kind of marks does a celluloid leave inside the lock?" Such an ignoramus was chief of security of a large and prominent hotel in New York. (Of course, 'loids leave no marks.)

True, many hotel men are familiar with known burglars specializing in hotels and have photos of them. But with a large influx of guests, especially during the summer and winter seasons (when such crimes reach their height) it's hard for security men to spot them. In European hotels and pensions, the concierge does act as a deterrent.

Those slick seasonal burglars

Much like Danny Sticks, the typical hotel thief is in his mid-thirties, smooth and glib. Blending into his environment, he's apt to play golf and tennis, hang around swimming pools to choose his mark. Often Danny checks in with an attractive female partner. He has a special liking for diamond jewelry, but everything from emeralds to rubies delights him. Not pearls: It's hard to distinguish the genuine from paste, so in a hurry he's apt to leave pearls behind.

Danny works fast, though not as speedily as one Arthur Lee Shank who probably holds an all-time record, breaking into thirteen rooms at one hotel within fifty-five minutes with his trusty picks. Most vacation-type burglars follow the sun, plying their profession at beach and mountain resorts in summer, at Florida and Caribbean hotels and motels in winter.

Danny stays at a hotel for no more than a week or two, then moves on to another area where he lays off at a resort, refraining from work until his next foray. Immediately after a score, he gets rid of the loot, usually by mailing it to a colleague for delivery to a fence. The fence then "breaks" the loot; prying the stones out of the identifying settings, melting down the gold-engraved charms of bracelets, rendering the loot anonymous. In many cases, the gold is sold for pennyweight value. If a really big score is made, the burglar will contact his receiver directly by phone. The fence will visit the resort, pick up the swag himself, quickly appraise it, and pay off on the spot.

Once, detectives walked into a New York hotel room and came up with the whole ball of wax. The burglar and the fence were sitting on the bed, the loot and a pile of money between them. The fence had a jeweler's loop attached to his eyeglass frames. The burglar's confederate had the cylinder out of their connecting room door, fashioning a master key for future work. Two phone numbers were found on that hotel burglar: his bail bondsman's and his fence's.

Danny's spree generally lasts for no more than two or three months, after which he quits for the remainder of the season. Unless something really big comes along.

You don't hear too much about rip-offs by thieves like Danny. Hotels tend to hush them up to avoid publicity. Hotel burglaries are more of an "impersonal" crime than the traumatic house or apartment theft. Hotel victims don't feel quite as shocked (they can go home to their nice unviolated home), and the hotel burglar is not likely to vandalize or leave any evidence of sexual motivation. Hotel thefts are generally the crime of a more calculating, if not more skilled, thief.

Most professionals preying on vacationers are pick men, though at times they'll 'loid a spring-latch door lock. Veteran Georgie "Irons" ("iron" is a burglar's term for a key) would loiter in a hotel lobby, scouting for names of likely victims from garrulous bellmen. Sometimes, for a worthwhile fee, a hotel clerk or maid revealed the name of a guest who had "specially nice" jewelry. Georgie would acquire room keys merely by lifting them from the front desk where they had been left on a busy afternoon, the clerk distracted. (Keys left at the front desk are for the hotel's convenience; fewer are lost this way.) Georgie particularly liked hotels that attached huge tags to the key. People hate to keep these with them when going out for a tour of the town.

Often Georgie worked blind, preferring corner rooms on upper floors, usually the most expensive. Both Georgie Irons and Danny Sticks, after staying at a hotel for a week or so, would "forget" to turn in their room key when they checked out. Later they'd return to burgle that room's future occupants.

When a pro has been a hotel guest for a while he may remove the cylinder from his front door, replacing it for an hour with another cylinder. With small locksmith files and blank keys he'll cut keys to various levels, trying the key in the door of an adjacent room, to produce the one to fit any door in the hotel.

Another ploy used by thieves is to ask a hotel or motel maid to open your door with her passkey, pretending it is their room; the thief may say that he had carelessly left his key with his wife, who is out shopping. When this ruse is used it is never the work of a real pro. He is violating a cardinal rule: In an eyeball-to-eyeball situation, he can be identified by the employee. And he is taking a hell of a chance that the employee does not know the occupant by sight.

A veteran burglar learns as much as he can about the security at the hotel he plans to hit. He gets to know about closed-circuit camera locations—and on what floors they're stationed. He knows the routes of scheduled "runs" of the security people in the hotel, the personality of those on duty and on standby; and what floors are favored with security. He knows that millionaires are apt to rate security checks every three or four hours. He needs only three or four minutes to pull a job.

Throughout Europe, hostelries will often be easier to burglarize. Still widely used there are the old "warded" locks with the classic keyhole in the door. Skeleton keys, good only for warded locks, are readily made. With a warded lock a guest on retiring may leave his key in the inside lock to block out other keys. But a daring thief can slip a newspaper under the door, push the key out where it would fall, then pull the paper out with the key.

Because these old-style locks are still found in Europe, a tool not seen in this country by generations of burglars is still utilized. It's called an *Ouisti*. Similar to needle-nosed pliers, it allows a thief to grab the front of the key through the outside keyhole and turn the key that has been put into the lock from the inside. Opening the locked door soundlessly, he proceeds to steal while the guests are asleep. The more modern European hotels now use pin cylinders; in good locks these have to be picked. Generally, if you stay in a modern hotel in Sweden, Switzerland, France, or England, you'll find pin-cylinder locks on your door.

Women burglars tend to be door shakers, prowling the corridors and trying doors to find one unlocked. Challenged by a hotel security man, they're likely to claim an "appointment" with a male guest. (So lock your door even if you're just stepping across the hall for a few minutes.)

Door shakers—male or female—are also inclined to be "pants burglars." They are thieves who want you to be in the room asleep while they commit their crimes, so that your most valued possessions are bound to be there.

Some women have been known to work with keys. In Miami Beach, police arrested a grandmotherly-looking thief whose car was crammed with room keys taken from hotels along the Strip.

People away on a fun holiday seem to turn garrulous. At a bar a congenial fellow guest may casually ask you, "Got any tickets for that new show? . . . Oh, we're seeing it, too. What night are you going?" Your acquaintance introduces himself and manages to get you to tell him who you are. Later, mentioning your name as he phones the hotel desk, he finds out your room number. The night you're enjoying the show, the thief enjoys breaking into your room.

Suppose, on another occasion, you return unexpectedly to your room (832) and find a stranger there. Chances are he'll pretend to be slightly drunk and stagger around authentically, demanding why *you* have intruded. You ask, "What room is this?" He'll reply, "Nine thirty-two." When you point out his error and ask how he got into your room, he'll say that he found the door open and thought he'd forgotten to lock it. Or the intruder may claim that he met a dame in the hotel bar who invited him up to her room—832 or was it 822?

On one occasion a security man caught a smooth would-be burglar warily roaming a corridor. The thief insisted he was looking for a Mr. Phillips. Suspicious, the security man moved to take him to the office and summon police. The burglar, highly indignant, whipped out a card. "I'm an attorney," he said, haughtily, "and I'll sue your hotel for false arrest!" Since hotels dislike such civil suits, the security man apologized without even glancing at the card.

Big-city hotel burglars like show-business personalities. The Lipman-Perenzo team studied *Variety* to learn who was coming to town, when and where they were to stay. Lipman would find out, through a string of hotel employees he cultivated, what suite was preferred by the celebrity. Then the pair rented that suite before the announced visit.

This was precisely the M.O. they relied on when they burgled Patrice Wymore, widow of actor Errol Flynn. For a week before her arrival at the Drake Hotel, Lipman and Perenzo occupied the suite she had reserved, hand-filed a key to duplicate the one for the door, and left the morning she checked in. The following night they hit the suite, coolly walking out with $8,000 worth of furs.

The motel burglar is usually not as skilled or as classy as the resort breed. With vulnerable key-in-lock doors he may use a jimmy or just force the knob with a vise grip. Except for rooms that open onto an interior corridor, most motel doors are visible to outsiders, so daytime break-ins are as unlikely as in hotels, where the dangerous hours for patrons are from 6:00 to 9:00 P.M.

The high visibility of motel doors, on the other hand, present the burglar with an opportunity he doesn't usually enjoy in a hotel. He can sit at his room window or in his car and watch the comings and goings of a large number of guests. From the kind of car you drive, he can evaluate whether your room is worth hitting, as he sees you drive off for the evening.

Just about every motor inn built in recent years has sliding glass doors opening out to patios or balconies. Some of these balconies are seemingly inaccessible from the outside. But the locks on these doors are not very adequate, making them a target for thieves or rapists. The man at the window, or parked in his car watching motel traffic, may not be a thief but a pervert intent more on the sexual than the economic target.

You may remember the shocking rape of singer Connie Francis by an intruder at the Howard Johnson Motor Lodge in Westbury, Long Island, New York, in November 1974. After having sung at the Westbury Music Fair, Miss Francis returned to her second-floor room and went to bed at 2:30 A.M. At 5:30 she was awakened by a young man who raped her at knife point. Tying her to a chair, with her hands behind her back, he knocked the chair over. She was then covered with a mattress and a suitcase. Her assailant left after taking a mink coat and some jewelry.

Police later determined that the burglar-rapist had taken advantage of a faulty lock in a sliding glass patio door. Charging that the company had failed to provide her with a safe and secure room, Miss Francis sued the motel chain for $6 million, was awarded $2.5 million by a federal district court, and settled out of court in 1977 for $1,475,000.

Small roadside motels will probably not have a security man on duty and they may not be too careful in their choice of employees. Not long ago the manager of a motor inn reported to police that eleven rooms had been broken into the night before with no apparent sign of forcible entry. All doors were fitted with cheap key-in-knob locks. The looted room occupants raised hell, contending that a key must have been used, and they accused motel employees.

Detective McDermott, called in to investigate, dismantled the locks and determined that the burglar (or burglars) had used a tool to force the locks. That tool tightly gripped the knobs without marring the surfaces. The forcing of these key-in-knobs by vise-grip pliers or a pipe wrench was run-of-the-mill burglary practice.

"But I figured," McDermott recalls, "that this fellow worked it so that no marks were left because he intended to revisit these rooms when new guests (and fresh loot) arrived."

Sure enough, about a week later that burglar did come back and two detectives collared him—an unemployed plumber's helper.

PRECAUTIONS TO TAKE

► If you're in an older European or American hotel and the key you've been given is one of those big, awkward bitted types made to fit a keyhole that you can see through, wedge the back of a chair under the knob when you retire for the night. Or use a good, hefty doorstop bought in your hometown hardware store. Some of these small door wedges have battery-operated alarms built into them, and they are a handy security investment at about $7.

► Take along that gem of a device, a portable travel lock, preferably the key-operated kind (see illustration). These fit easily into the edge of a hotel or motel door and can't be opened from the outside. Made by several manufacturers in the United States, they are lightweight, small, and inexpensive. You may also use such a lock, when you leave your room, to secure the top drawer in a dresser or night table containing valuables not deposited in the hotel safe. If there's a connecting room, place the travel lock on your side of the door.

► Bring with you on the trip only jewelry you are certain to wear; leave the rest behind in a bank safe-deposit box. The jewelry taken along should be kept in the hotel's vault and you can make your selection for the evening, returning it the following morning.

► To scare off intruders when you are out of the room, consider a portable alarm device. Best for this purpose is an ultrasonic, space-protecting contraption. (See Chapter 11 for more information.) Plug it in on the far side of the room, facing the entrance. Since it has a thirty-second time-delay feature, when you enter there's time to deactivate the alarm with a key. A burglar unaware of the alarm's presence and unable to shut it off will certainly run when it sounds.

► Suppose you find yourself unprepared, with neither travel lock nor alarm, and there's no sliding bolt or short chain bolt on the door. How to keep the burglar from walking in on you? Use whatever is at hand to block the door. You might tilt your luggage between the floor and the doorknob, or pull over the night table and tilt it against the door, if you have to. If your lock is picked, or a key used, the intruder will either be unable to move the door at all or make enough noise in doing it to awaken you and give you time to shout and phone for help.

► Never leave cameras, binoculars, or other expensive items lying around on a table or bureau top where they can tempt hotel workers or even other guests passing your open door while the maid is cleaning up. Keep these valuables locked in your luggage. If possible, place the luggage in a locked closet (or one on which you've placed a travel lock). Keep at home a list of the articles' serial and model numbers.

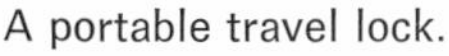

A portable travel lock.

► Leaving your room, use your key to lock the door if possible. If the maid has already made up the room, hang a Do Not Disturb sign on the outer doorknob.

► Keep your key with you during the day and evening, rather than dropping it off at the hotel desk. If the clerk doesn't know you, a slick thief who has you down as a mark may see your key in the rack and just step up and ask for it by room number. He has been loitering in the lobby so his face may be familiar to the clerk, who assumes he's a guest.

► To fend off door shakers, simply keep your door locked at all times.

► When out of your room, leave the hotel or your portable radio playing. Out at night? Leave your lights on as well.

► Never reveal your room number to casual acquaintances at poolside or at a bar. Keep your key in your pocket or purse; those big digits on the tag can be seen yards away. If you look like you're going to be away from your room for a while, your room number is all a pick man needs to know.

► Don't reveal your planned excursions to other guests.

► Of course, use traveler's checks and/or credit cards instead of cash. Reminder: In a different pocket or purse keep an up-to-the-minute list of checks paid out, in case the remaining checks are lost or stolen and you must report it.

► If you leave children alone in the room, tell them to turn the inside lock turnpiece when you leave. In a number of hotels this should cause an "occupancy" indicator to show from the hall and it will deadlock the door.

► Deposit your return travel tickets in the hotel vault.

► Men traveling alone should be reminded that prostitutes with "hearts of gold" are a myth. And the "nice girl" so lonely in the cocktail lounge may not be so nice at all or so lonely.

► A woman traveling alone? That "nice guy" may be on the prowl for an easy, gullible mark.

► Try the patio-door lock in your hotel or motel. Be sure it works properly before accepting the room. Are you nervous about a ground floor or other accessible location? Insist on another room.

► Realizing that you don't have too much personal control over your security when traveling, take out some form of insurance. The policy should have riders for theft (cameras, jewelry, furs, and luggage) that will provide coverage during your trip or vacation.

What hotels are doing

The Connie Francis rape has shocked major hotels into beefing up their security. The Hyatt hotel chain, for instance, uses both chain locks and security guards, and some of its larger hotels now have peepholes and closed-circuit TV surveillance.

You may have heard about "no key" hotels. Several new electronic systems have been designed to replace the conventional lock and key. One, Cardgard, functions with a wallet-sized plastic code-punched card inserted into a small panel (*keyport*) at the door. A similar control card is "read" by a console behind the hotel's front desk that signals the code to unlock the door. If a card is reported stolen or lost, the desk clerk removes the control card from the console and a new card and code are issued. Thus far, few hotels have taken to the Cardgard system in the United States, none in Europe.

With another new no-key system, Digikey, the guest punches digits in a door panel as on a Touch-Tone phone. Each guest chooses his own door-opening combination, which is fed into a front-lobby control panel. At least eight Hilton hotels in the United States have installed Digikey.

The difficulty experienced in some of these systems can be attributed to the hotels' requiring access to rooms for maid service.

With the more complicated Telebeam system, when a guest checks in he's given a plastic card with his key. The combination of holes punched in the card is unique to that card and with each guest the combination is changed. After the guest uses the key to enter the room, a buzzer sounds to warn him he has twenty seconds to insert the card in a slot at a terminal on top of the TV set. If the card isn't put in, an indication appears at a manned console. When that happens a roving security man is sent to the room. That was how, on one occasion, a holdup of a group of jewelers (accosted as they entered their suite) was foiled.

Another time, a sales executive attending a national conference in a New York hotel remarked in a crowded elevator that he didn't use his card and the buzzer didn't sound off. He had called the desk and was told the alarm in his room was malfunctioning. "What the hell," he groused in the elevator, "nobody around here cares."

It turned out someone did care about his alarm. That evening while the executive was at his dinner meeting, someone—undoubtedly a person who had been in the elevator listening to him—picked his room lock and stole his expensive camera, gold cuff links, and travel alarm clock.

An increasing number of hotels and motels have been installing room safes or strongboxes in rooms, to protect valuables that guests cannot or will not place in a vault. Many of these safes and boxes have keys. One type, TelSafe, is equipped with a lid as well as a key. Only when the key is turned and withdrawn can the lid be locked in place on the strongbox. Even if a thief makes a duplicate of the key, his chances of finding the mating lid is remote, since the next guest to use the room will be handed a different lid and key.

Various hotels are testing ever new security measures. With the Lok-a-Wat, an arriving guest unlocks his door, walks into his hotel room, turns on the light and TV, and wearily flops down on his bed to relax. Two and a half minutes later, the TV, the lights, and air conditioner go off. To get the electricity working again, he simply throws the dead bolt in the lock on his door—thus making himself more secure.

The Lok-a-Wat consists of a metal box that fits under the desk in each hotel room. Wires connect it to the lock, TV, lights, air conditioner, and heater. Two and a half minutes after the guest leaves the room, everything is turned off.

This insures your use of the protective bolt on your door when you are in your room, but you can never hope to fake out a potential thief in these places while you are out. With the Lok-a-Wat you can't leave a radio playing and a light on when you go out. The device conserves energy but does nothing for your security once you leave the room. Places that use Lok-a-Wat are concerned less with burglary than with utility bills.

Hotel security men, reluctant to depend on such devices, contend they are still unproven and there have been some failures in operation. Many prefer lock cylinders like Emhart's new High Security system, which is extremely resistant to picking and whose keys are hard to duplicate. (See Chapter 11 for more information.)

Let's take a brief look at one supreme hotel thief and his brazen operations.

"Raffles" of the hotels

Clifford Daniel Monroe, baby-faced and curly-haired, was a dapper, remarkably nervy twenty-two-year-old specialist in hotel thievery. A former spot-welder, he couldn't pick locks; so his only burglary tool was a 'loid, usually a Do Not Disturb sign he removed from a doorknob in whatever hotel he was invading. He bought such show business publications as the *Celebrity Bulletin* to learn where well-heeled and jewel-laden Hollywood visitors would be staying.

One evening Monroe left his girl friend as she was stepping into her bath at their West Side apartment in Manhattan. He hailed a cab to go to a posh Fifth Avenue hotel, where he walked confidently past the doorman. An elevator took him to his "lucky" ninth floor, where he appropriated a Do Not Disturb piece of plastic. He 'loided two other doors but found nothing worth stealing. Taking the stairway to the tenth floor he came upon a corner door marked 1007–11. To Cliff, the multiple numbers meant it was a suite, an indication of occupants with money. After ringing the bell four times and receiving no response, Cliff slipped his 'loid in the door and entered.

In the dressers he pounced on fifty-four pieces of jewelry, which he wrapped in two handkerchiefs. Cliff knew it was the largest score he had ever made. Departing, he slammed the door shut behind him. On the street the doorman summoned a cab for him. (Later, Monroe was to remark to detectives that he felt embarrassed because he could not tip the doorman; his pockets were so stuffed with jewelry that he was unable to get at his money.) He arrived at his apartment just as his roommate was stepping out of her bath. That's how fast Cliff Monroe worked.

He presented his girl with the best of his loot, which turned out to be the 20.23 carat emerald-cut World's Fair Tiffany diamond. There were a few phone calls to fences. After a viewing, a deal was made for $44,500 on the whole lot, though Monroe actually received only $7,000 for his thirty-five minute excursion. Newspapers reported the true value of the gems stolen that evening at $364,000. The *New York Times* referred to the thief as a "Raffles."

Cliff's victim turned out to be Mrs. Irene Selznick, widow of a Hollywood producer and now herself involved in theater production. She had been out that evening at the premiere of a highly publicized new movie. Having been once burglarized in Paris, she had had her own lock put on the New York hotel door—but on the evening of the premiere she failed to use her key to double-lock the door. Among the jewels Mrs. Selznick had left lying around—besides the Tiffany diamond, valued at $130,000—were a bracelet containing 15 pear-shaped, 20 marquis, and 50 baguette diamonds. And a $200,000 brooch.

A "squeal" by another burglar who also used Cliff's fence led detectives to tail Monroe and he was nabbed in a subsequent break-in. In the detectives' squad room he was thoroughly interrogated, flattered with the encomium, "You're the Maury Wills of burglars." He was told the police knew he had done the Selznick job and had been badly cheated by his fence, Eddie the Jeweler.

"You might as well tell us all about it," a detective suggested, "and we'll go after that bastard Eddie. That's who we want most."

So Cliff Monroe, incensed at the way he had been "robbed," opened up and talked. And talked.

He checked off fifty-three hotels, each of which he had burglarized at least once and some as often as five times. He had worked regularly about every night, sometimes committing two burglaries the same evening. Eventually Monroe admitted to more than a hundred jobs within a year, for a total haul worth nearly a million dollars. Although he "earned" more than $90,000 during his yearlong spree, gambling, high living, nightclubbing, and show girls left him virtually penniless. Indicted on twelve counts of grand larceny and burglary, he was handed a sentence of three to seven years.

As for the victimized Mrs. Selznick, all she would have had to do to prevent the loss of her $364,000 in jewels was to *place them in the hotel vault—or double-lock her door when she left for that premiere.*

Chapter 6 Making business and industrial premises more secure

UNIFORMS can be impressive—and deceptive. One morning in April 1974, a tall young man walked into the Paramount Jewelry Exchange in lower Manhattan. The Exchange, where some fifty merchants maintained booths, boasted extensive, highly sophisticated burglar alarm systems. A police station was located only a block away. The man wore a blue uniform cap bearing the insignia of the Holmes Protection company, which provided the alarm systems, and a windbreaker with a similar identifying shoulder patch. Dangling from his belt were pliers, tape, and other instruments. A black toolbox in his hand also displayed the Holmes insignia. He made his way directly to the control equipment near two large vaults.

One of the jewelers, observing the stranger, called the attention of a security officer always on the floor of the diamond treasure house. He sauntered over to the young man and casually asked what he was doing.

"I'm a repairman from Holmes," was the cool response. "Our company has been having some trouble in the past few nights with false alarms in this neighborhood. We think maybe it's the truck traffic. So I've been sent to check your alarms, to make sure they're in good condition."

The security officer, satisfied, walked away.

And the uniformed "repairman" proceeded to alter the alarm control circuits so that they would be inoperative but still permit a closing signal to work when the vaults were closed—ostensibly assuring that all was well in the system.

Shortly after the Jewelry Exchange closed at 6:30 that night, five men entered the place from an adjoining building, forcing a basement door. Opening the wooden alarmed doors of the cabinet built around the safes, the burglars were confronted by two thick steel doors operated by combination locks. They burned these with oxyacetylene torches. Another set of metal doors was forced, to expose, finally, about a hundred safe-deposit boxes.

Those boxes were filled with diamonds, gold, jewels, and cash deposited by the member merchants. In all, the gang's haul that night was estimated at $3 million.

Another gang victimizing jewelers was not nearly as refined. Unable to defeat an alarm system by sophisticated means, they were nevertheless successful in getting what they were after. Many jewelers, putting all their security eggs in one basket, were relying on a very stable and dependable Holmes system. They would deposit all their valuables in concrete or cinder-block vaults and turn on their "pressure-vault system." This type of system monitors the air pressure within the vault; any variation caused by an opening of the vault door or a breaching of the walls will trigger an alarm.

Operating for many months, a group of burglars known to the police as the Pressure Vault Mob wreaked havoc throughout New York's jewelry fraternity. Entering a building full of jewelry manufacturers just before the close of the business day, one member of the mob would secrete himself. After closing he would admit his confederates, armed with sledgehammers, crowbars, and pickaxes. They would chain all the entrance doors shut, plug the cylinders of the outside locks with chewing gum or liquid solder, and then make their way to the upper floors of the building. Elevators would be immobilized and all doors behind them chained and nailed shut. They never did a job beneath the sixth floor. Holes were battered in walls leading from the fire stairs. A jeweler's plant would be entered without causing any alarm; they never made a break on a door because they were afraid the door might be protected (wired).

Once outside, they'd sledgehammer their way through the concrete-block vault wall. As soon as the wall was penetrated the alarm would be received at the protection company. Armed company guards, as well as police, would respond—only to be met by the barricaded doors. In the upper reaches of the building the burglars would keep on smashing and pounding until they could climb through the opening in the vault wall and get at the treasure of gold and precious stones. Making their way to the roof, they used ropes to swing to adjoining buildings, removed the ropes behind them, and hid anywhere in the buildings on that block. After the frustrated Holmes men and police had left, the burglars would emerge and vanish.

Those astonishingly adroit burglars, the Pressure Vault Mob, were an aggregation of some twenty men working in five-man crews. They had pulled at least thirty-five jobs in the past year, their average loot $35,000 on each job. The police, under pressure from the Jewelers Security Alliance, assigned four two-men Safe and Loft teams on roving patrols in the jewelry business sector.

One night the following month word came through on the police radio that an alarm-protected jewelry concern had been hit. The regular precinct police and Holmes men rushed over, found the break-in, took notes for their reports, and left the area. The four special police roving units, however, hid out, surrounding that block. Two hours later, five Mob men carrying their loot strolled out of a nearby building—into the waiting guns of the police. One of the burglars in this crew turned out to be the Mob's ringleader. After that "collar," pressure-vault burglaries ceased.

Comments Detective John Kid, currently Robert McDermott's successor as the outstanding burglary expert in New York's Police Department:

"To prevent such break-ins, jewelry merchants should have perimeter protection—alarm systems on every door, window, and other means of entry. They should also have space-protecting ultrasonic or microwave systems to warn of an entry before the actual attack on a vault or safe. Safes and vaults should be of the type that resist sledgehammers and certainly resist the punch, drill, and rip kind of attack. The Pressure Vault Mob never hit places with space protection simply because they would not have had time to crack a safe."

And, we might add, the Jewelry Exchange alarm defeat would have been short-circuited if the security officer there had telephoned Holmes Protection to confirm the alleged repairman's identity and the reason for his visit.

A short time after the successful hit at the Paramount Jewelry Exchange, two men identifying themselves as detectives entered a jewelry store in Queens, New York. They told the owner they were from the local police squad; in view of the many burglaries at jewelry shops, they wanted to check his protection. As the men opened the control panel for the shop's alarm system, the proprietor asked if they'd mind if he phoned the precinct to verify their purpose. The men immediately left the store.

What disturbed the shop's owner, he later reported, was the fact that these "detectives" had a key to his locked control panel. Otherwise he probably would have welcomed the attention the "police" were giving his security.

The business burglars

There are thieves who specialize in breaking into office buildings for cash, stocks and bonds, or whatever else of value can be stuffed into pockets or brief cases. A certain type focuses on cash registers, safes, vaults, and safe-deposit boxes. And commercial establishments are plagued by the business burglar who targets in on fur or camera stores, jewelry or coin shops. He'll break a front window with a brick, grab watches, diamond rings, or tape recorders, and run. He is known as the "three-minute" or "smash-and-grab" burglar.

Some of the daring three-minute breed, like Marty Krone, go for premises that are "protected" by silent alarm systems that signal an alarm at the monitoring service office. Marty knows that he'll have no more than five minutes in which to seize his swag and scoot, so he times himself for a three-or-four-minute hit. Having previously reconnoitered the place carefully, he knows just what he wants to pick up. Marty will ignore a shop that does not show an alarm system emblem; he figures the pickings there are too slim to warrant his effort. If his target is a camera or hi-fi store, for instance, he knows where the most costly items are displayed and grabs them at once.

Loot taken by these thieves is sold for a fraction of its original value. Much of it is gotten rid of by the burglars themselves in the city's poorer sections, around bars and factories, and to lower-middle-class suburbanites struggling with time payments and mortgages. Portable TV sets, particularly color models, are popular.

A "cat" burglar is primarily interested in racks of furs, bolts of cloth, cases of expensive whiskey. Usually short and agile, he often wriggles through air ducts and skylight panels or through transoms. He rarely works alone. His task, usually, is to get inside the building without making an obvious break and then open a rear door for the rest of his gang.

Like others in their trade, commercial burglars are habit-prone and conservative. They will almost never give up hitting stores to turn their attention to apartments or houses.

One variety of industrial safe burglar specializes in substantial companies or plants where many workers receive their wages in cash. Construction firms especially are of this type. Here, the thieves strike on the night before payday. Greedy, not content with cash only, they may look for checkbooks, slipping out some blank checks from the back of the book so they won't be noticed until later. Then the burglars peddle the blanks to an expert forger.

Another favorite target of the "smash and grab" species is the quality men's clothing store. Parking two or three cars at the curbside just in front of the store, they will force the front door and enter. Most of these stores have glass doors, the lock engaging at the bottom of the door into the door saddle or sill. By placing a tire iron under the door and lifting, the bolt can be removed almost entirely out of its strike (the recess in the sill). Then, hitting the door with a shoulder, they are in. The alarm has flashed to the central office of the alarm company, the cavalry is on its way. The burglars, rushing in, scoop up as many expensive suits as they can carry from the racks, run out, and dump them into the waiting cars, which speed off before alarm guards or police can reach the scene.

Detective Robert McDermott offered a simple suggestion to clothiers through their trade associations: Hang the garments on racks with the hangers facing alternate directions. Burglaries attempted right after this advice was heeded proved its effectiveness: The thieves could no longer just grab off armloads of two-hundred-dollar suits. They had no time to make worthwhile scores, and the practice fell off.

Then there's the thief who specializes in cash registers. Arthur Reilly, an aging pick man so crippled by arthritis that he walked with a slow, painful shuffle, plagued midtown Manhattan for several years. In the early evening, while there were still plenty of people on the streets, Arthur would pick the lock of an alarm-protected retail store and go for the locked cash register. No, Arthur did not try to pick the register open; he'd merely lift it off the counter and drop it to the floor, letting it hit on a back corner. Sure enough, the drawer would fly open, he'd scoop up the cash and walk out. By the time the radio cars arrived he was at least a block away.

People depending on highly sophisticated alarm systems frequently neglect the essentials, such as a really good lock on their door. Taking advantage of this failing, an imaginative pair of old-timers made their fortune by burglarizing fur lofts. Lurking on fire stairs, they watched employees file out of their selected victim's premises and then saw the owner diligently use his keys on the three or four ordinary door locks. Before the furrier had reached the street the two thieves would be at the door with picks.

Virtually every perimeter alarm is actuated only on actual intrusion; that is, the alarm doesn't sound until a door is opened or a window moved. This larcenous team, after picking the locks on the door, quickly moved in, stealing as many furs as they could carry. In two or three minutes they were out. They reset all the locks, which they had left in a "picked" position, by merely swinging the keyway in the opposite direction. They were on the street long before arriving police or guards.

The alarm company personnel, accompanied by patrolmen, would find the door intact, all locks secured. Entering, they'd find nothing untoward and depart. The owner would be welcomed on his arrival home with a report from the protection company that an alarm had been received, his premises had been checked—and the incident was classified as a "cause not found." False alarms are not uncommon. Not until the furrier opened his loft next day was the burglary discovered.

"Monkey Joe" had still another approach to fame and fortune. Joe worked from rooftops, coming down the sheer face of business buildings on a rope. He reckoned, and wisely, that very few businessmen will go to the expense of placing alarms on windows opening to nothingness. So down he came, counting the floors until he was suspended outside the victim's area. Then he broke a window, reached in, unlatched it, and went after merchandise.

Joe liked expensive women's apparel, specializing during the season in the latest fashion lines about to be introduced. (He had a market among the victim's competitors.) Joe packed the clothing in cartons, threw them out of a rear window, and retrieved them later from a backyard.

Occasionally Monkey Joe's buyer was located in the same building as the victim. In this case Joe wouldn't have to go to the trouble of throwing the loot down to the courtyard; he'd pile it up at the protected door, open it from inside and hustle the swag into his buyer's loft. Joe would be snug with his loot there, behind those locked doors all night, while police and alarm company guards went through the building and eventually departed empty-handed.

How you are vulnerable

Locks on office and merchant doors have the same weaknesses as those encountered in homes and apartments. Heavy store doors don't deter a thief if the locks can be picked or forced with a jimmy. A glass skylight on a shop or factory may be unprotected. Important keys to a door or cabinet, left lying around during the day, may be "borrowed" temporarily by a dishonest employee, duplicated, and quickly returned. The copied keys can then be sold to a professional crook.

The combination to a safe may be known to an employee who quits his job or is fired. Or, if you leave your safe open during the day an employee can take off the inside combination lock cover by removing a couple of screws and read the combination right off the tumblers.

Before a burglar takes out his hammers and drills, he searches around. He knows that bookkeepers, secretaries, and owners file the combination numbers under "S" for Safe, "N" for numbers, "D" for Dial, or "M" for Money. They may write the combination numbers on a piece of paper and tape it under the pullout leaf of the desk. It may be written on the underside of the wall calendar, other places familiar to the career safe burglar.

You are vulnerable, too, when you are opening your safe. All a thief needs is to observe this operation occasionally and eventually he will have the three numbers he needs; the fourth is the opening number and he knows it will be between 90 and 10.

In closing up a large shop, office, plant or department store at night, managers may neglect to see that all windows and doors are firmly bolted or locked. Often, too, no search is made to be sure that a stranger isn't secreted in a toilet, basement, broom closet, stairwell, or other hiding place.

Remember this: If the alarm will not set properly, the employees or owner often leave anyway, assuming it is just a closing trouble, not realizing that the condition may have been caused by a burglar. The burglar will wait to see if the alarm is repaired and set.

Particularly attractive to thieves are any openings in the rear or on the roofs of buildings. One-story shops in suburban areas usually offer window and roof openings for attack. Hatches in roofs should be fitted with eyebolts and secured from inside with heavy-duty padlocks. Bars should be installed beneath skylights. Delivery doors should be protected by barricading with a metal or heavy wood crossbar.

You are susceptible to a later burglary attack if you neglect to observe good security during your open periods. One burglar was able to make a key and negate the alarm system of a travel agency so that his pals could make an easy entry later that night. This job was possible because the thief was able to remove the lock cylinder in seconds—the screws that held the cylinder in place were right where he could get at them. It would have been even easier for him if he could have taken the door right out of its protective circuit by getting at the wired contact above the door; then he could just have removed the wires from the contact screws and twisted them together. As it was, all the burglar was able to do was slice through the ribbon of silver foil on the door, causing the alarm to sound as the agency attempted to close. The owners did what was expected: They left before the system was repaired, believing it to be only a "closing trouble."

Merchants often clutter front windows with so many posters or signs that a patrolman on his rounds can't see into the store if a burglar is at work there. Too often, retailers neglect to empty cash registers before leaving at the day's end. And at night and over a weekend or holiday, a shop may have inadequate lighting inside and at exits and entrances.

Where and how they break in

Burglars enter business premises through all kinds of openings—basements or roofs, delivery chutes, elevator shafts, air vents, coal chutes, side doors, and rear windows, whatever is most easily penetrated. At warehouses and supermarkets it can be through a rear loading platform or wooden door where a small drill and keyhole saw cut a panel from the door. Certain thieves are fond of breaking in by removing air conditioners and exhaust fans. Others like to work with an inside accomplice, an employee who lets them in after hours, having first disconnected a local alarm system.

When a safe burglar is ready to go to work, the first thing he will do is try the safe's door handle. If that doesn't work, he looks to see what position the dial is in. Knowing that your combination probably requires that the last number you dial in unlocking is somewhere between 90 and 10 on the dial, he puts slight pressure on the handle and moves the dial *back* one digit at a time. Very often just doing this will open the safe.

Why does this work? If you merely spin your dial *once* when locking, you may not have sufficiently dispersed the tumblers. Because the vast majority of safes have a fixed opening number (one that can't be changed) and that number is within the 90-to-100 and 0-to-10 range (twenty digits) he has a fair chance.

The employee whose job it is to open your safe every morning often tries to save himself some time, just moving the dial away from the "opener" when he closes the safe. Then all he will have to do the following morning, he reasons, is move it *back* a few digits and he won't have to bother with all that manipulation.

"After all," he rationalizes, "I'm the only one who knows the final number, and anyone fooling with the safe is only going to scramble the numbers further. So why not do it the easy way?"

The average, common safe, designed primarily to safeguard

cash and valuable papers from damage by fire, will not resist a determined attack by even an unskilled thief. With tools normally carried in a car trunk, such a safe is easily defeated within fifteen minutes.

The safe can be *punched,* that is, the numbered dial hammered off, the exposed dial spindle driven into the safe with a steel punch. Many a burglar relies on *peeling,* or *ripping*; with this technique a chisel or sledgehammer is used to spring open a top corner of the door. Then a crowbar is inserted into the opening and worked down, peeling off the steel layer until the boltwork can be reached.

Most safes today are constructed more lightly than in the past, using lock spindles designed to thwart the punch but not the rip technique. So burglars have taken up the drill method, attacking the units with a portable electric drill at a specific point of vulnerability. Their goal is to circumvent the newly designed "punchproof" locks by punching through a hole drilled in the front plate. The method demands a great deal of skill.

Safes are vulnerable to several other methods, including oxy-acetylene "burn" technique. This technique requires extensive equipment: burning or cutting torches, tanks of oxygen and acetylene, protective glasses, striker, tips, tip cleaners, insulated gloves, and more (see photo). It's rarely necessary, however, for a burglar to resort to burning to defeat a safe; he's more likely to use it on chests, which are more secure and are specifically designed to safeguard property from theft. (For more on safes and chests, and suggested choices, see Chapter 11.)

A chest will yield only to certain skilled burglarious techniques. Punching and peeling doesn't work; and even burning is difficult because of copper laminates and other materials used to disperse the heat of the torch. Still, some chests will succumb to the torch when the burn is directed to surfaces other than the front door. Chests may also be burgled by drilling, using special bits and framework to allow for exceptional pressure.

As you know by now, burglars tend to hit on a particular M.O. and stay with it. One quartet of thieves specializing in safe burglaries concentrated on their favorite victims, Chinese restaurants—and with good reason. They found the cash scores

Oxyacetylene tanks and tools, with various weapons, used by safecrackers.

from these safes much higher than elsewhere. Apparently these particular restaurant owners let a great deal of money accumulate before making a bank deposit.

Six or seven of these jobs had occurred before detectives uncovered who the thieves were. One member of the mob was an Englishman named Parker, an employee in an electronic manufacturing firm; another, a black, was Yarbo, from Jamaica; the third, a New York skin-popper (mild heroin addict) named Marty Frank; and the fourth was Phil Rosen, our former itinerant photographer. In ripping off the restaurants, the mob was addicted to roof breaks. Despite the weakness of a door lock, for instance, they would come in through the roof. If there was no skylight they would use brace and bit to cut a hole in the roof and then operate a keyhole saw to make an opening. Then they lowered themselves on ropes to get at the safe. There, they employed only one method, the rip, to get their score. Newspapers reporting their activities described the group as "master burglars" and "yegg-men" (safeblowers).

One of the burglars was an informant of Detective McDermott's, who relates what took place:

He asked me to meet him one night at Broadway and Ninety-third Street. Sitting on a bench I saw him walking toward me with a peculiar gait. When he came closer I saw his problem: He was wearing a pair of shoes probably a size and a half too small for him; I assumed they were loot.

He was angry. A couple of nights before, he was out with the guys who were doing the Chinese restaurant jobs. During the day they had gone from one restaurant to another looking them over, and he was stuffed with egg rolls before he thought one was worth doing a job in. It had no alarm system and the place appeared to be a cinch. On leaving with his pals, he told them it looked bad, figuring he would do the job himself.

"What the hell," he told me. "I can rip a safe as well as the next guy."

After visiting a few more places the quartet broke up, indicating a lost night. My informant picked up a bag of tools and returned to the restaurant he'd chosen. While inside ripping the safe he heard a noise in the rear. Looking out of a window he was shocked to see his three partners climbing up to the roof. He called out, "Hey, what are you guys doing?" They ran.

"Screw them," the stoolie told me. "They were going to do the job without me. So here's their names and the plate number of the car they use when they're out for a score. I know, I know what you're thinking, I'm double-crossing them. But this, this was the *three* of them against me; I'm going to drop out of the thing and you can collar them."

So we busted them about two weeks later. They had gone after a Chinese restaurant in White Plains, outside New York. Going through a skylight they startled two employees who slept in the place, tied them up, and ripped open the safe—for $400. I was surprised to discover my informant among them—he had not "dropped out" yet. His only comment to me: "Gee, Mac, I thought it would take you a little longer to get them lined up." All four were sentenced to two and a half to five years.

So much for honor among thieves and the intelligence of the average safe burglar.

PRECAUTIONS TO TAKE

Few people realize that real security is rarely attained by dependence on a single device, whether it's a door lock or a barred window. Effective security demands an intelligent amalgam of antiburglary devices, sensible precautions, and vigilance commensurate with conditions at specific premises. In nonresidential quarters these are some of the main rules to observe:

► Permit only trustworthy employees to handle keys or to know your safe's combination.

► The thief is looking for the "soft spots"—the favorable conditions. In your home you know you can rely on your family not to "cooperate" with a burglar; unfortunately, this is not always the case in your business. Aside from examining your premises for physical vulnerability, be wary of the information and access given employees.

► Install solidly built, closely spaced metal bars or steel mesh beneath skylights.

► When you are buying a lock, be there when the locksmith

installs it and get the keys from him yourself. Has an important key disappeared? Change the lock cylinders.

► Protect air conditioners and exhaust fans with steel bars and place a meshed screen over the bars. Have openings wired by an alarm company.

► Most security experts agree that the best single deterrent to commercial burglary is a good alarm system. If you are in a business dealing in merchandise attractive to the thief, use the alarm company in your area that offers the best form of line security to prevent sophisticated defeat of the system. (See Chapter 11 for specific choices.)

► Install an alarm system at such points of entry as delivery chutes and doors in elevator shafts. Use infrared light beams to crisscross open floor areas or aisles.

► Never leave your premises for the night with your alarm "in trouble." Wait for the alarm-repair response.

► Do not put in any kind of device designed to trap a criminal on the premises.

► Bolt down expensive equipment such as cash registers, electric typewriters, and other office equipment, using keyed locks made for the purpose. These will allow you to move equipment from one desk to another when necessary.

► Unused windows should be bricked up. Fit other windows with padlocks or bar them with heavy shutters or folding gates that don't allow tampering from the outside.

► Replace thin wood-paneled or hollow-cored doors with metal-clad or solid wood doors. Make sure outside hinges are not removable and use nonretractable screws on double-cylinder locks. Choose locks that resist both picks and jimmies. Doors to loading platforms should be made of steel, not wood, and must be locked from the outside. Sheet steel should cover basement doors on both sides.

► See that lighting at rear doors as well as at recessed doorways is adequate.

► Guard against easy access from utility poles, alleys, roofs, fire escapes, and loading docks.

► Closing your shop at night, take costly items out of your display windows. Slide folding metal gates across the front and secure with a pick-resistant padlock. Be sure all windows and doors, including the basement door, are locked securely. Empty your cash register and keep it open at night. Have some lighting throughout the store. Keep a light on, especially near your safe if it's visible from the entrance or through a window. Don't fail to set your alarm system. Make bank deposits before the store closes.

► Your safe should be anchored to the floor, placed where it can be seen from the street. A chest with a round or square door is preferable. Never write down the combination and leave it around, or set it to your birth date, phone number, or address. (For more information on buying a safe, see Chapter 11.)

► Do not keep a gun in your safe. Very few safe burglars carry a gun on the job, but if you do have a gun there they will probably steal it. Nor will they hesitate to take on a night watchman if the loot is worthwhile; so don't provide the weapon.

► In checking your alarmed premises before leaving at night, look at the contacts above your door. Make certain the wires are attached to the two screw terminals, that they are not removed and twisted together. (If the wires have been tampered with, you may still get a "good" closing signal, tricking you into thinking the alarm is working.)

► Make sure the set screw holding your door lock cylinder in place is not accessible. Buy a lock that protects this screw with an armored face plate.

► Secure all perimeter doors from inside with padlocks and bars except the one through which you must leave. It's not enough just to protect your vault.

► A good lock cylinder will not only resist picking, but will also make impossible any drilling and shimming, or making of a key by impression. One good lock that protects against these assaults is better than several ordinary ones.

► Never use an externally mounted on-off (electric shunt) switch for your alarm system. Use a time delay to allow you to get inside without the alarm sounding—and then use a key in your alarm control box.

► Notify police beforehand if you'll be handling a large sum of money on special occasions.

► Make sure repairmen are legitimate.

The case of the incurable safecracker

Some years ago two Wall Street brokerage houses reported the mysterious loss of almost $2 million in securities over a period of four months. Several midtown retail stores also informed police that cash was missing from safes. In all these crimes there was no evidence of forcible entry into the premises and the safes, vaults, and chests showed no marks of attack. Oddly, in every instance some money was left behind.

Store proprietors and office managers concluded that one or more employees was involved and several of them who had keys or access to the combinations were fired. Three long-standing partnerships in stores were dissolved, based on the belief that one of the partners had been tempted to steal.

When McDermott was called in to investigate he removed the lock cylinders from all of the doors, and his minute examination disclosed a tiny hole in the face of the lock cylinders. The hole was covered by a filler consisting of wax and gold leaf. That discovery eliminated the dishonest employee theory, and uncovered an entirely new method of burglary.

The rash of these odd safe-burglaries prompted the Safe and Loft Squad to assign four detectives to patrol Fifth Avenue, where several stores had been hit. Early one Saturday evening a detective's attention was drawn to a tall, well-dressed man in his fifties, strolling along apparently aimlessly. What attracted the detective was the man's casual alertness as he seemed to window-shop.

At one corner the man paused before a shoe store to study the window display there. Then he moved toward the recessed front door out of line of sight. What especially piqued the detective's curiosity now was the fact that the store sold only women's shoes. From across the street the detective observed his man peering into the dimly lighted store interior. Aware that his "subject" might see him reflected in the window, the detective moved along. A few minutes later, crossing the street, the detective realized he had somehow lost his man. The subject had vanished.

Now all four detectives on patrol converged on the scene. Police on surveillance duty have a habit of assigning a nickname to subjects being tailed, usually based on obvious physical characteristics. That night's subject had been dubbed "Willy." Since Willy had not appeared on the street after being observed near the shoe-store door, they felt he was now inside. That meant they had to wait for him to come out.

About twenty-five minutes later Willy reappeared, paused momentarily outside the door, and then walked, apparently bemused, into the stream of Fifth Avenue pedestrians. Surrounding him, the detectives searched him, then placed him under arrest.

In the man's pockets they found $910 in cash. Plus a jeweler's pin vise, a #75 drill bit, a standard hand pick, a small envelope containing wax and gold leaf, several slivers of fine metal (watch springs), a small flashlight, and light cotton gloves (see photo). When the store's proprietor was notified and came to the store, he reported that $897 out of a total of $1,200 (the Saturday receipts) was missing from his otherwise "undisturbed" safe.

The safecracker turned out to be William O'Rourk (the nick-

Willy O'Rourk's tools of his trade included tension bar (beneath pliers); jeweler's pin vises (to make a minute hole in cylinder); tube of liquid solder (to hide hole); crank-shaped wire to remove chain bolts; blank keys to lift pins out of cylinder; pieces of gummed paper labels to paste above dial in safe.

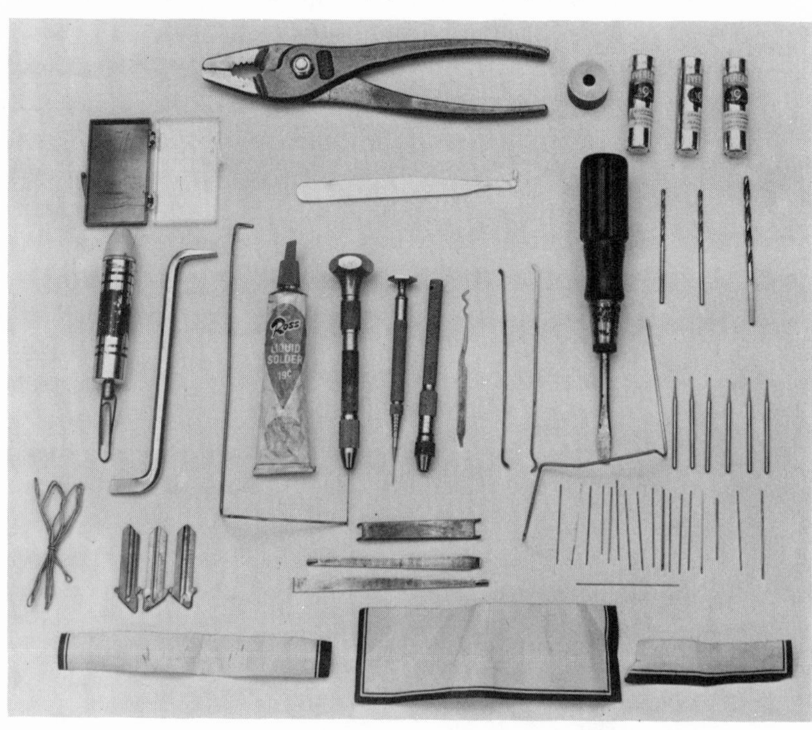

name "Willy" had been a fortuitous coincidence). According to his statement, O'Rourk was a skilled machinist with his own machine tool company. On failing to meet his company payroll some months before, he had taken to crime. O'Rourk had long been intrigued by safe manipulation as a kind of "intellectual exercise." When his back was up against a wall financially he put his know-how to work.

It was Willy O'Rourk who invented the drill-and-shim technique, now commonplace with burglars. Drilling a minute hole in the face of the lock cylinder just above the keyhole, he would then slip a watch spring (shim) into the hole. Using a pick he lifted the cylinder pins until the shim separated the pins, thus opening the lock. But it was not just this new method that was so remarkable. O'Rourk was absolutely the greatest safe manipulator the New York police had ever encountered.

Before his appearance on the scene, detectives investigating losses from safes had discounted the possibility of manipulation without prior knowledge of the combination. Yet here stood Willy, a mild, middle-aged man with tired eyes, surrounded by detectives, pleading guilty to burglaries that insurance companies had written off as "mysterious losses . . . internal thefts."

As his almost incredible talents were recounted to a judge, O'Rourk agreed to cooperate with the police. He directed detectives to the offices, stores, and brokerage houses he had invaded; the stolen securities were recovered and the three broken partnerships mended.

Willy, claiming he wanted to "contribute materially to the welfare of the community," turned out to be a gold mine of information for safe manufacturers as well. The Safe and Loft Squad arranged for the major safe companies to send representatives to meetings with Willy at police headquarters. These meetings were chaired by Robert Murray of the Murray Safe Company, the best of the legitimate safe men. (It was Murray whom the federal government had sent to Europe after VE Day, to open by manipulation the safes of Hitler and other members of the Nazi hierarchy.) Willy advised the manufacturers on how they could make combination locks that would "keep me out." They listened, and as a result safes now have locks classified as MR (manipulation-resistant) and MP (manipulation-proof).

Willy was handed a ten-to-twenty-year sentence, but in view of his cooperation it was suspended and he was placed on probation to continue his consulting with the safe makers.

Six weeks after he was given his big break, the Safe and Loft Squad, doubting his sudden conversion to the side of the angels, nabbed O'Rourk again while he was doing a job. So off he went to prison. Released aften ten years, Willy went to work for a burglar-alarm company. An old man by then, he dutifully performed his duties every day. Occasionally McDermott watched him for a day or two, once observing that "Willy walks with the prison gait . . . he is scarcely aware of his surroundings."

But Willy's fingers still itched, and one night he returned to the same Fifth Avenue shoe store he had burgled years before. The report to Safe and Loft read: "Cash missing from a safe. . . . No force. . . . Method of entry not readily apparent." McDermott

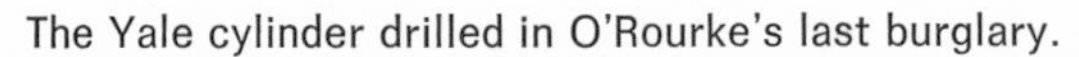

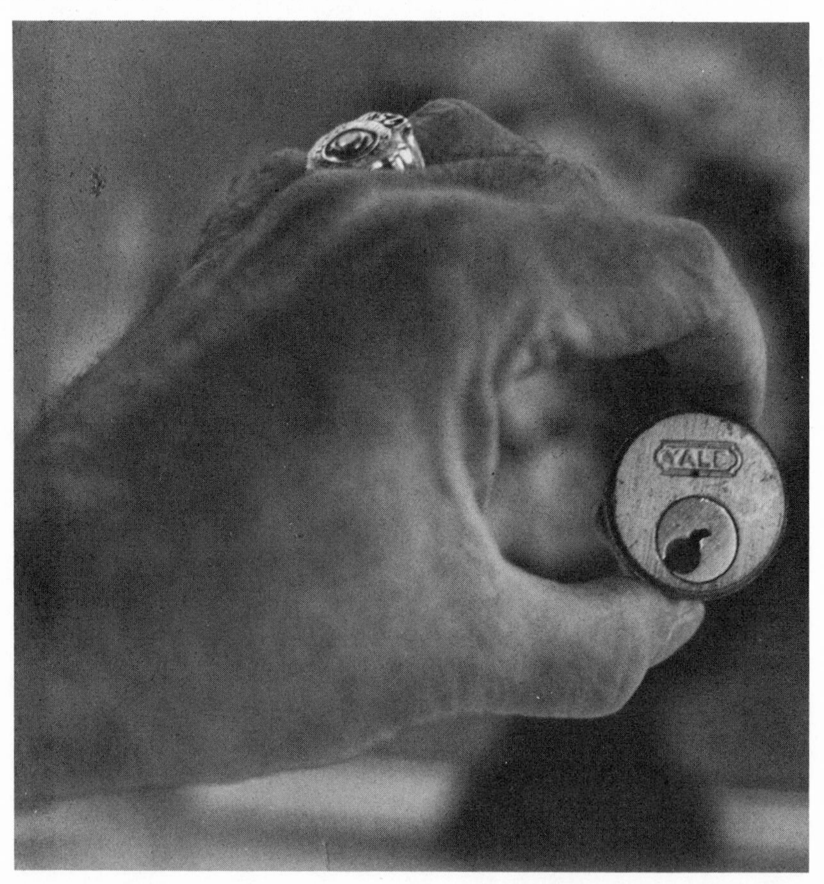

The Yale cylinder drilled in O'Rourke's last burglary.

again found the small hole in the front-door cylinder. This time the hole had been obliterated with a gold crayon and it was not in the same spot; it was a half-inch higher (see photo). Incurable Willy couldn't stop.

Two Safe and Loft detectives were put on O'Rourk's tail around the clock. A week later he was again arrested, this time as he came out of a midtown office building at 11:00 P.M. In one pocket of his jacket Willy had $427 in cash; in his threadbare topcoat pocket, a pin vise, pick, and shim. And a slender book of poetry entitled *The Men That God Made Mad.*

One of the young detectives who doubted Willy's genius asked him how he had obtained the safe's combination.

"Listen, kid," Willy responded, "on the fourth floor there's a locked safe with a pair of men's socks in it. On the fifth floor there's a safe with a bottle of twelve-year-old Chivas Regal. I made the score on the eighth floor. The rest of this building must be on the verge of bankruptcy."

After this bit of bravado O'Rourk added: "I owe ten years—I'll die in jail."

That night O'Rourk, as a parole violator, was on a bus headed for Sing Sing.

There is a postscript to the story. Three years later McDermott received a letter from a thief named David Berg, an inmate at Sing Sing. Davy, doing two-and-a-half to five years for burglary, had been put away for stealing half a million dollars in jewelry over a fourteen-month period. He wrote Mac that he had a new gimmick, learned from "an old jerk named Willy."

"You just drill a hole in a cylinder," Davy wrote, "and use a watch spring from a ladies' watch . . ."

Incidentally, Berg signed his letter, "Houdini." In a news story covering Davy's arrest, a reporter had referred to him as a Houdini and the thief never forgot it. At the time his only tool was a celluloid strip, yet he had called O'Rourk "an old jerk."

Chapter 7 If you encounter an intruder

IT HAPPENED in a two-bedroom, third-floor apartment of a nine-story building on an August morning in a still fashionable upper East Side neighborhood in Manhattan. Many of the building's tenants were professional and career women. Among them were twenty-two-year-old Janice Wylie, a striking blonde, ambitious to be an actress, who worked at a *Newsweek* clipping desk. She shared the apartment with Emily Hoffert, twenty-three, waiting for a job as a teacher; and Patricia Tolles, twenty-one, an apprentice researcher at *Time* magazine.

That morning Janice had planned to take the day off, telling friends she was going to Washington for a civil rights protest march. Apparently she changed her mind. She slept late—in the nude, as she usually did in warm weather. Emily, preparing to move to a new apartment, left with some luggage and said she'd return shortly. At 9:30, Patricia went off to work.

Pat returned home at 6:30. Unlocking the door she walked across the entrance foyer, took one step into the living room, and stood rooted. Lamps and chairs were overturned, bric-a-brac scattered over the floor. Pat called out for the others. Receiving no response, she fled to the apartment of Max Wylie, Janice's father, a well-known writer, who lived a few blocks away.

Mr. and Mrs. Wylie, with Pat, stepped into Janice's bedroom. It was a shambles: clothes strewn about, dresser drawers pulled open. On the floor was Janice's nude body. She had been stabbed so savagely that her intestines were torn out of her stomach; and there were seven knife wounds in her heart. Smears on her body, and an open jar of Noxzema on the night table implied that a sex maniac had molested her.

Emily's body, fully dressed, lay on top of Janice's. The women were obscenely bound together with strips of white cloth, the wrists of each woman tied to the ankles of the other. Emily had been knifed around her neck, her jugular severed. Below the sink in the bathroom was a foot-long carving knife, streaked with blood. A second knife, the blade snapped at the handle, had been dropped in the bedroom.

Signs of a desperate struggle were evident. The scene looked as though a wild animal had ripped and torn through the room. The wall above the women's bodies was streaked with blood. On the floor, an empty soda bottle somehow made the scene even more macabre.

The rear service door leading into the kitchen was ajar about two feet, apparently how the killer made his exit. Pat recalled that she had slammed it shut in the morning when she put out the garbage, had turned the bolt and double-locked the door.

Police checked the building's two entrances leading to the courtyard. One detective observed that the women's kitchen window was open from the bottom. No bloody fingerprints were found; evidently the killer had washed his hands before leaving. Nothing of significant value was missing from the apartment.

From the outset disagreement prevailed among the detectives studying the carnage. Some felt this was the work of a burglar who killed to prevent being identified. Most were convinced that the savagery of the attack on Janice indicated a strong personal motivation, certainly sexual and with suggestive lesbian overtones.

Obviously Janice had not been attacked as she slept. A raincoat that she normally used as a dressing gown was found on the floor in the apartment's foyer. She must, it was reasoned, have taken it from the foot of her bed to cover her nakedness when she was awakened. To answer a ringing doorbell? No forcible entry was evident and those subscribing to the personal motive theory argued that Janice must have admitted her assailant. Others held that the young woman had probably been awakened by a noise in the apartment—a noise made by an intruder. Clearly, Emily Hoffert had interrupted the attack on her roommate and was slain to silence her.

The Manhattan North Homicide Squad needed more than speculation and the Safe and Loft Squad was asked to have

Detective Robert McDermott look at the locks. This was the substance of his report and his further recollections:

"The locks from the murder apartment were brought to the Safe and Loft Squad and I was called at home to examine them. It was about three o'clock in the morning when I got to headquarters. I found that the lock cylinder hadn't been picked and that no master key existed. The lock was of the type that could be left open if a button was depressed on the lock case. When I got the locks the button was set which would allow the lock to be opened from outside by just turning the knob.

"Under the microscope I found scratch marks on the latch bolt and they were of very recent origin. These I knew to have been made by a metal instrument used to depress the spring latch, much as a celluloid strip is used.

"At about five-fifteen in the morning I went to the apartment. The strike plate set in the frame of the door showed identical scratches, and I was certain that someone had made a very recent entry using a shim stock of brass or steel to depress the latch, opening a door that could not have been double-locked. The report I wrote that night concluded:

" 'It is the opinion of the assigned, based on this examination of lock case, cylinder, and crime scene, that a metallic instrument, thin and supple, having a sharp edge, was used to dress the spring latch and effect an entry. The cylinder was not picked. Shim stock of brass or steel is sometimes used by burglars instead of the usual celluloid. A venetian blind slat is commonly used.' "

McDermott recalls the unforgettable scene.

"After examining the front door I went through the entire apartment. A patrolman was sitting in the living room keeping his silent vigil. The place was otherwise empty. I saw chalk marks, flashbulbs, and fingerprint powder everywhere. In the kitchen, a half case of empty soda bottles. The window was open fourteen inches. Looking out the window I saw a sheet-metal ventilation duct protruding about two and a half inches from the brick wall just under the kitchen window. About five feet away and below the level of the apartment window I could see a wide open fire-stair window. After checking the service entrance door into the kitchen I remembered thinking that if it weren't for the minute marks on the front-door lock I would say the killer came in through the window.

"Three days later, Captain Weldon of the Manhattan North Homicide Squad called me. 'You were exactly right, Mac,' he said. 'A venetian blind slat was used. One of my men told me that three days before the killing a locksmith changed their locks and he used a venetian blind slat to show them they had better double-lock their door.' "

With that logical explanation for the marks on the lock, detectives again pursued the theory that Janice knew and admitted the killer.

McDermott still felt, though, that a burglar was responsible; with all else eliminated, the man must have come in through the kitchen window found open when detectives first viewed the apartment. Why a burglar? Evidence by now had been unearthed that Janice had received obscene and vaguely threatening phone calls at her office. McDermott placed a lot of importance on the raincoat found in the foyer. She had *not* put it on. If she were answering a doorbell, Janice would have worn it but she had not. She had grabbed it and clutched it to her as she was frightened awake; she must have stepped out of the bedroom and met the intruder in the hallway. To McDermott, the dropped coat meant she had been struck by the soda bottle right on the spot. The bottle came from the kitchen—the kitchen with its open window.

The Wylie-Hoffert murders spurred one of the most massive manhunts in the city's history. The yearlong investigation involved hundreds of suspects, thousands of false leads. The constant news coverage kept the public at a fever pitch and, like the earlier Boston Strangler murders, had the women of the city terrified.

Suddenly the tension was relaxed. Brooklyn detectives announced that nineteen-year-old George Whitmore, Jr., arrested for rape, had confessed to the Wylie-Hoffert murders. He was charged and indicted.

Officially the Wylie-Hoffert case had been closed, but on their own time some of the detectives were still on it.

Months later, while Whitmore was still awaiting trial, the case was finally really solved with the arrest of another man, Richard "Ricky" Robles.

It turned out that a certain drug pusher and his wife who had been picked up by police weeks before had told police that on the night of the murders a customer of theirs had admitted to them that he had killed the two young women. The customer (Robles) had asked the drug pusher and his wife if semen in the throat of one of the girls could be (chemically) traced to him. Up to that time, this semen factor in the crime had never been revealed to the public. Besides the detectives actively assigned to the case, only the killer knew exactly what had happened that fatal morning.

The drug pushers, hoping their cooperation would help them when they came to trial, agreed to work with police. After their apartment was bugged, the couple induced Robles to give them fuller details of his crime.

"I grabbed the window sill from the stairwell window below," he blurted out, "and lifted myself up to the kitchen. . . . I went to pull a lousy burglary and wound up killing two girls. . . ."

As Janice came out of the bedroom, the panicky Robles attacked her with a bottle he had picked up as he passed through the kitchen. Then the "sex factor" came into play. He sexually assaulted her orally and then, snatching up a jar of what he thought was Vaseline, smeared her body and his penis with Noxzema. He complained that it "burned" him.

"Then," Robles went on, "the other one walked in—she walked right in."

From the bedroom doorway Emily had called out: "What's going on here? Oh, I see, I see what you're doing." Robles grabbed her.

"Leave my glasses alone," she said, enraged. "I want to be able to identify you."

Robles hit her with the soda bottle. Deciding to kill them both, "I went to the kitchen for a knife."

(George Whitmore, recanting his confession, claimed it was made under duress; three witnesses placed him out of town that day, and charges against him were dismissed.)

At Robles's trial he was found guilty on two counts of first-degree murder and sentenced to life imprisonment. One juror later remarked:

"The girl who came in, the schoolteacher [Emily], really told him off. If she had acted differently he never would have done this."

"If she had acted differently . . ." How differently? What would *you* have done? What should you do if and when you encounter an intruder in your home or office? What's the etiquette of behavior while being burglarized?

In the case of the career women's confrontation with their intruder, certainly Emily should not have announced she wanted to be able to identify him later. That served to goad Robles to his final act of violence, to make sure he would not be recognized and picked up by police. The instant Emily noticed the living room was in disarray she should have left, gone to a neighbor, and summoned police.

As for Janice, if she had been awakened when she heard an intruder moving about in the apartment she would have been wise to use the phone at her bedside to quietly contact the police (dialing Emergency 911)—and then call out, "Who's there?" If she had opened her eyes to see a strange man already in her bedroom, Janice should have feigned sleep, letting him burglarize the place and depart. Unfortunately, she evidently heard Robles coming in through the kitchen window and went to investigate the sounds, screaming only when he was already almost upon her.

In very few cases is an intruder intent on inflicting physical harm, sexual or otherwise. It was only after he "had to hit her" that Robles became sexually aroused. If Janice had *immediately dialed the police and then called out,* there's little question but that Robles would have fled through the kitchen door before the tragic confrontation.

PRECAUTIONS TO TAKE

Whether you live in a private house or in an apartment, here are some cardinal rules to observe:

► Approaching your house, you notice the door is partly ajar, damaged, or unlocked, or a window is open that you're sure was shut when you left home. Or a strange car is parked in your driveway. If you observe any of these things, a burglary may be in progress. Rather than entering your house and tackling a thief or thieves, go to a nearby neighbor and phone the police.

► If you come face to face with the intruder while you're inside your residence, don't be a hero. Don't threaten or tackle him. If you can get out quickly, do so. If not, try to stay calm. You might say, "I won't fight you or try to look at you. Take what you want and go." Generally he's after cash, watches, jewelry. Chances are, he's scared himself and wants to get out of the situation. Don't panic him. But do make mental notes of what he looks like.

► If he's already in your bedroom and you wake up, don't scream. Feign sleep. You may be dealing with an unstable personality like Ricky Robles. If you yell he may react violently to quiet you.

► Never get between the burglar and his exit. He'll turn violent, inflict harm as he makes his getaway.

► If you believe an intruder has broken in but you don't know where he is, never turn on lights near you or call out until you have phoned the police emergency number. Then let him know you're aware of him, that you are at home. Block or lock doors between you.

► Don't use a flashlight because this spots you to the criminal. If you have to rely on a flashlight, hold it away from you at arm's length. Don't target yourself in case he has a weapon.

► If you hear a prowler coming into your residence, make a noise such as kicking a wall, and shout. The average burglar is sensitive to such noises and he'll climb right back out of a window or wherever else he is making an entry.

► Whenever possible, get out of the house and to a neighbor where you can summon police.

It's possible—and not too uncommon—to be hoodwinked by a smooth, glib burglar you may suddenly come upon in your residence. Take the talented Freddie Wallace, surprised in an upper-middle-class midtown flat. The new building was still not fully tenanted and some construction workers roamed about. Appropriately, Freddie wore a workshirt and chinos. While picking up valuables in an apartment he heard someone at the door and ducked into the bathroom. As the tenant, a lissome divorcee, walked in, Freddie called out, "I'm in here" and emerged, much to her alarm.

"Tile man," Freddie smiled, holding up a pencil and piece of paper. "Just checking to see that your bathroom tiles are okay. We're finishing up on the building."

A mild-looking, handsome young thief with an infectious grin, Freddie got into conversation with the lonely woman and was invited to stay for a drink.

At the end of a half hour Freddie departed. The woman, adjusting the blinds at a window, saw him walk quickly down the street, step into an Oldsmobile beside another man at the wheel, and drive off. Only several hours later did she discover that $7,000 in jewelry was missing. Warning: Beware of strange visitors.

Should you keep a gun at home?

If you have any reluctance about using a weapon, don't keep firearms in your house or apartment. The average citizen does not have the skill, or enough control, to frustrate a potential assault with firearms. The intruder could attack, turn your gun against you. *Threatened violence begets violence.*

A gun can only *possibly* give you comfort if you are at home when an intruder is present. Otherwise, when you're out, the thief would probably find it and arm himself with it; then if you interrupt him your weapon becomes his. The only way to make sure your home is your castle is to keep the criminal out.

Generally the law says you can use "reasonable" force in defense of your own well-being and property. But the interpretation of what's regarded as "reasonable" may vary from one judge and jury to another. As a rule, you have the right to employ physical force only if the intruder is actually breaking into, or is already inside, your residence.

If you shoot an intruder, you may find that the courts in your state hold that you, in your own home, have no more right to do it than do police who shoot a felon on the street. In fact, you may have fewer rights. For example, if you use a weapon (even a gun you possess legally) to stop a burglar fleeing from your home, *you* have technically committed a crime.

Here's a rule of thumb: You can use only force commensurate

with the force or threat of force being used against you. So if you do shoot a thief he had better be inside your home—and his wound in the front of his body, indicating that he was coming at you. A policeman on the street (or elsewhere) is required to abide by pretty much the same guideline, except that he can shoot a "fleeing felon" if the cop has reason to believe that if the criminal got away he would be a physical threat to others.

If a woman is attacked, and she jabs her assailant with a hatpin, she can be prosecuted if the degree of force used to repel the attack exceeded the force employed against her. If the force she used was equal to that of her attacker—but she later states that "I carry the hatpin for protection"—she will not be prosecuted for the assault. However, she may be indicted for wielding a dangerous instrument. The law is peculiar in this regard.

Defend yourself, yes, but be sure the "weapon" you employ is "normally" carried on your person or in your purse.

State criminal laws vary, but you may well be subject to *civil* suit even if the law in your area says it's legal to carry certain "deterrent" weapons, such as a tear-gas projector.

If you have a gun, you're unlikely to use it on a dangerous intruder—and if you do, he may shoot back. In the words of one attorney in a western state, "an honest man with a firearm has about as much chance against a criminal as a jackrabbit would have at a coyote convention." Yet the chances of someone being shot *accidentally* are high. The "someone" could be a member of your family, a neighbor, or a friend. Guns in the house are often used in family quarrels and can be too handy for potential suicide by a deeply depressed individual. Every year about 1,200 Americans are killed and more than 10,000 gravely injured in shooting accidents in the home.

Suppose you hide the gun under lock and key. Then it won't be easily accessible to you when danger arises. Unless you've had plenty of practice, you probably couldn't hit a burglar at five yards. And if you came upon an invader who has a gun himself, trying to shoot him would involve you in a gun duel.

What about a tear-gas projector or substances like Mace for protection, to spray a thief? Sure, it could repel an attacker but the chemical can also enrage him and make him far more violent. And you could get a bad dose of your own poison. Such devices are illegal in a number of states, including California, Illinois, and New York. Check your own state law.

Remember, your goal is to keep criminals *out* of your home. Having a gun, Mace, a baseball bat, or other weapon around the house may give you a false sense of safety—and make you ignore or overlook basic security measures such as effective locks and alarm systems.

And yet, entirely defenseless—and imprudent—you could turn out to be the victim of an unstable burglar/killer.

The shooter: murder by a climber

Common burglars are generally labeled cowards, fearful of a victim offering resistance, and unlikely to invade occupied premises. They are also unlikely to carry a weapon themselves. Among the exceptions are "climbers," invariably window thieves lacking lock-picking skills. Many are peeping toms. The very nature of their modus operandi compels them to work under cover of darkness, on the blind. This makes them dangerous. Quite often, unintentionally, they enter dwellings where someone happens to be present. A jittery armed climber, surprised on a job, can kill.

Shortly after 10:00 P.M. one balmy spring evening, all was quiet at a neatly kept apartment building in a middle-class neighborhood of Flushing, Queens. In the ground-floor hallway James Braddock, the affable superintendent, ambled toward his own apartment carrying a six-pack of beer presented by a grateful tenant whose stove he had just repaired. Along the way was a vacant apartment, shown to prospective tenants by Braddock's wife earlier that evening. Presumably a noise made the super pause there.

Using his key, Braddock opened the door. Evidently he heard a noise, turned on the light switch, faced a stranger, and moved to tackle him. A shot rang out, the bullet hitting Braddock in the chest. He stumbled back into the hallway and collapsed to the floor, dead.

Detectives summoned to the scene later found no evidence of a break-in. They theorized that Braddock, accosted in the hallway, had been forced to admit the man to the apartment where the super was shot. A ski mask, found in a rear courtyard and evidently abandoned by the killer as he fled, pointed to a window as his avenue of escape. No motive was immediately apparent.

Detective Robert McDermott was assigned to the case. Here is his story.

At that time I was in Canada, lecturing before the World Congress of Police Officers. When I returned to New York two days later, I was asked to determine just how the murder happened.

At the building I was told that Mrs. Braddock, on showing the vacant apartment, had made sure the double-hung wooden inner windows were securely locked. But I detected a small hole in the screen of an aluminum storm window. I figured the intruder had used an ingenious technique.

First, with a small tool he pierced the unlocked screen to lift it out of his way. Then he introduced a screwdriver between the upper and lower sash of the window close to the latch. Taking advantage of the side play between the window and its frame, he rotated the tool to move the latch out of its strike.

The technique was so unusual that it must have been the work of a burglar who had applied it often before—probably someone who had graduated from using a jimmy. If the window had been loose enough in its frame he would not have needed any tool to defeat the window lock. I submitted my report and the windows were removed to be photographed. Police investigators now knew they could eliminate anyone having keys, or access to keys, for the apartment. And that Braddock had not been met at gunpoint in the corridor by a stickup man who forced him to open the apartment door.

Specifically assigned to the puzzling case were Detectives Frank Moore and Charles Prestia of the Queens Homicide Squad. Moore and I had worked closely on a couple of cases before. Most recently a housewife, Mrs. Rabin, had been found stabbed to death, a telephone in her hand, in the kitchen of her ransacked home. Chief of Detectives Fred Lussen had called on me to find out whether there had been a burglarious entry or whether she had admitted the killer or killers to the house.

Examining the locks on the three doors in the Rabin home I had been unable to find any scratch marks, any sign of forcible entry. Nor was force evident on any of the windows. Moreover, all doors and windows were protected by an alarm system. I reported that in my judgment, despite the ransacked upstairs bedrooms, a burglary had *not* been committed. "The victim," I added, "had admitted the perpetrator(s) by depressing the button of the alarm to allow entry without sounding the alarm."

Throughout the house it seemed that a little earthquake had struck: dresser drawers had been tumbled haphazardly on the beds and floor, the contents of living room cabinets scattered around. But while articles of trivial value were missing, expensive items had been ignored. It all reinforced my conviction that a burglary had been simulated by a killer or killers known to Mrs. Rabin.

Within a couple of days two young men were arrested. One, Jack Brown, had recently been fired from his job at the Rabins' dress shop. The other was a friend of his, a soldier based at nearby Fort Totten. In Brown's confession he related that he and his buddy had gone to see Mrs. Rabin to ask for "some money." At the entrance the soldier had stood to one side when Brown rang the doorbell. When she let Brown in and then saw the soldier with him, Mrs. Rabin tried to slam the door shut. They pushed their way in, demanded money.

Threatening to call the police, she ran to a phone in the kitchen. One of the men grabbed a knife and stabbed her. She fell on her back. Straddling her, the killer plunged the knife into her again and again. The two men then put on the "burglary" act and fled.

Now, at lunch with the detectives, we discussed the quick wrap-up of the Rabin homicide. Frank Moore recalled that the police had originally thought that that crime was committed by burglars.

"Mac," said Charlie Prestia, "from what you know about this Braddock case, this time it was really a burglar, right?"

I nodded. "Sure. Someone did come through the window recently. The marks are fresh and the flat was just painted to get a new tenant. But let's consider a few other things first. Wasn't the gun a twenty-five-caliber automatic?"

"We found two casings."

"Okay, not many flat burglars carry guns. A twenty-five, a

lady's gun. And this was a vacant flat. Do you think Braddock was meeting a woman there? What have you found out about him?"

"Everything good," Moore said. "We thought about his maybe going to meet somebody in the vacant flat. But when he was fixing that stove for a tenant, he acted like he had nothing on his mind. When the couple gave him the six-pack he stood talking with them at the door. And his wife is only down the hall from the flat. No, the guy is clear."

"And his reputation is good," Charlie added. "He's well liked."

I asked: "No trouble with former tenants? If the ski mask is connected, and this was a premeditated killing, it might be because Braddock knew the guy. . . ."

Moore shook his head. "Or we have a burglar who works at night. He can't be sure people aren't going to be home or walk in on him. So he always wears a mask."

"I feel sure," I said, "it's a burglar. I want us to get away from the 'unknown man.' "

We went over every bit of information, looking for anything that would take away from the "casual-motive" murder—the impromptu, random killing that is the bane of a detective's life.

"Why did Braddock go into that apartment?" Charlie persisted. "If he had heard anything when he went by the door, wouldn't he have left the six-pack on the floor in the hall when he went in? The beer was found inside. And from what Mrs. Braddock told us, he was not an alky. Going to drink or stash the beer in the vacant flat so his wife wouldn't know. Besides, we know he had beer in his refrigerator. Nope, I think Braddock was passing the door, through for the night, and just went in to check that everything was right."

"That's it," I said. "It's an unknown man—the thing we don't want."

Finally I came around to a course of action.

"Listen," I said, "that guy has an unusual M.O. and I think we can use it. We need to look for other burglaries by him. He is strictly windows. He may or may not always wear a mask. He works at night—he's a climber and he carries a gun."

"What did he have to shoot Braddock for?" Frank demanded. "The poor guy was getting old, and whoever did it must be pretty young, in good shape. He could have just belted Braddock and gotten away."

"Because," Charlie said, shrewdly, "he wants to hurt people, and I'll bet he's used a gun before. Maybe even that gun."

We decided on an unorthodox game plan. Since we knew the killer's offbeat M.O., we would trace his past burglaries. We also agreed on a name for our quarry, "the Shooter."

The two Homicide men and I adjourned to the 109th Precinct, which was to be headquarters for the investigation. When we unveiled our plan to Joseph Coyle, Chief of Queens Detectives, he was enthusiastic. How much time could I give to the case? Calling the Safe and Loft Squad, I was able to clear myself for a week or so.

We started by assigning a young police trainee to the task of combing reports of burglaries in the 109th Precinct. He was to spot all burglaries in which the method of entry was unknown; where windows may have been used; where windows were broken close to the latch; and any reports of people seeing a prowler. Our best bet, we felt, was to find a complainant in a burglary committed by our man, a victim who had supplied the police with a good enough description of missing property so that in recovering it we could pin down the identity of the person disposing of it—the Shooter.

By 9:00 P.M. that day Charlie Prestia's hunch that our man had used a gun before, maybe even this gun, was confirmed. Ballistics telephoned the 109th to report that in processing the Braddock death bullet they had decided that the same .25-caliber automatic had been used to shoot two other persons.

The first shooting occurred in the Bronx about six weeks before the Braddock murder. A tired elderly man, driving his car, had pulled out of traffic to rest and had fallen asleep. Awakened by a stickup man, the old gentleman offered no resistance, handed over his wallet. The bandit started to walk away, returned to the car, and deliberately fired a bullet into him. Surviving his wound, the victim described a masked young man, white, under six feet tall, slender and well dressed.

The second known use of the .25 happened a week later in Queens. A young teacher, Mary Kane, living alone in a Thornton

Place apartment house, walked from her living room into her bedroom to retire for the night. A young masked man suddenly pounced out of a closet and grabbed at her. In her nightclothes she ran screaming out into the hallway, pursued by the intruder. He fired a shot after the fleeing woman, hit her in the hip, and escaped through a window.

The third shooting of course, was ours, and he had killed. From what we now knew, it was clear there would be a fourth.

Early the next morning Frank, Charlie, and I drove to Thornton Place to examine the windows in the young woman's apartment. Understandably, she had moved after her unnerving experience. Other detectives were to interview her where she worked.

On the way over, Prestia commented angrily: "This guy isn't just a shooter, he's a homicidal maniac! He shoots the old guy in the Bronx for no reason—and he keeps the gun. He sure as hell shot this girl—and he keeps the gun. He shot and *killed* Braddock, and I'll bet he keeps the gun."

I thought to myself, he keeps the gun because he can't bear to part with it. But he's making it easier for us in the end. I pointed out: "If what we're doing comes out right for us and we get a fix on him there'll be no long 'inquiries' like where were you on the night of . . . Because when he's taken he'll have the gun in his pocket—or in his hand."

The marks on the Thornton Place window were almost identical to those in Braddock's vacant flat. It had been the Shooter. The gun wasn't being handed around among burglars. It was *his* work. We were right. We could track him.

We drove away from that apartment feeling we knew exactly what kind of a person the Shooter was, that we could even describe him.

"He'll be young," said Frank, "and I think tall, yes, from his reach on the screen and the angle of the shot at Braddock, tall."

"Wears good clothes," I added. "Neat but flamboyant, with a flair. You know how I know? Because this one is cocky, he's one of the *invincibles*. And the first one of us who comes near him with a gun in hand is going to be blown away."

Back at the 109th squad room we were told that the interview with the young woman victim at her job had produced only a vague description of the attacker. Nothing had been taken from her apartment.

That night I took home with me the .25-caliber burglary reports pulled from the 109th Precinct files by our police trainee. They went back over a period of four months. During the next eight days Charlie, Frank, and I visited about two dozen of the victims. We found two burglaries that I attributed to the Shooter. True to form, the descriptions of stolen property on these two jobs were virtually useless. On the ninth day we came upon a third possibility.

Just three days before the Braddock murder a young couple (we'll call them the Johnsons) returned late from work to their Flushing apartment to find it torn apart. The window leading from a fire escape was wide open, the glass smashed. Examining the double-hung windows I detected the distinctive marks. The windows, closed and locked, fit tightly in the frames and our climber had been forced to break the glass to release the latch—his unique M.O. failing him.

Mrs. Johnson had reported her oboe stolen. The description of the musical instrument, valued at $600, was excellent. Our Property Recovery Squad was immediately contacted to start a canvass of every pawnshop, secondhand dealer, and musical instrument outlet in the city. From the dealer who had sold Mrs. Johnson the oboe we got a photograph of an identical instrument, and copies were distributed to every command in the Police Department.

Four nights later, after again looking at burglary scenes all day, Charlie Prestia and I returned to talk with the Johnsons again. Detailing what was being done to locate the oboe, we pressed them for anything else they could tell us. "Did you get any phone calls just before the burglary? . . . Wrong numbers or the caller hanging up without speaking? . . . Were any of your wife's underclothes missing—how about the hamper? . . . Was the toilet flushed when you discovered the burglary?" (Burglars, nervous on a job, often do not flush because they fear the noise will be heard by a neighbor who may know the residents are not at home.)

The Johnsons, obviously baffled by our persistence, were

getting more nervous by the minute. We left after two hours, assuring them we would continue the search for the missing oboe. As we waited for the elevator, Charlie muttered, "Something stinks."

Next morning a message awaited me in the 109th squad room, asking me to call Mr. Johnson at his office. It turned out that after Charlie and I had left them the night before, Johnson had phoned his mother. She reminded him that the oboe had been left at her house.

"I'm sorry about this," said Johnson, sheepishly. "The thing was not stolen at all."

Of course, a claim had been filed for insurance. At this point, disgusted, I turned the telephone over to the day-duty detective. Frank Moore notified the guy's insurance company. Burglary is not the only "crime of opportunity."

Now we were plodding along, running out of prospects but still confident our scheme could work. Chief Coyle, hanging in with us, put out an order that every detective command in Queens was to pull all burglary complaints and list those with window breaks into apartments. Our young trainee was told to gather information on crimes committed since the Braddock killing. I went back to Safe and Loft, working the homicide only on an occasional morning or evening.

Then came the night when Frank, Charlie, and I, ever hopeful, were visiting Mr. and Mrs. Abe Breamer. By this time we had pinpointed five burglaries attributed to the climber, all leading nowhere. The Breamer's apartment, situated just a few blocks from the Braddock building, was to be the sixth.

At 11:00 I stood in a drenching downpour on the Breamer's fire escape with a small penlight in my hand. I was so anxious to get back into the dry apartment that I nearly missed the two small trademarks on the windows. The Breamers had been burglarized eight days after Braddock's murder. Their list of property stolen, discouragingly, gave no serial numbers. When Mrs. Breamer offered us coffee we all accepted with alacrity.

Frank read off the few items stolen and routinely asked if anything else had been missed since their complaint was filed. The answer was equally casual.

"Well, yes, I think so," Mrs. Breamer murmured. "We have a Shell gasoline credit card. Two of them. I never use the one Abe gave me so I didn't miss it right away. I'm not even sure it isn't still in the house someplace. I thought it was in my dresser drawer and I noticed just today that it's not there."

Our stunned silence must have been overwhelming.

The Breamers sat rigid, curious. None of the people we were interviewing knew we were on a murder case. To them this was solely a burglary, important only because it was theirs. Yet our interest in it was obviously more than routine. We couldn't hide our excitement over this new development.

"I bet our guy took the credit card," I said, "and he'll use it or sell it. Either way we've got our handle now. Can you imagine his moxie, doing a job within shouting distance of the kill?"

We helped Mrs. Breamer search her dresser drawers again. Not a sign of the credit card. Her husband showed us his duplicate card and produced gas-station receipts indicating he had used it only once, in New Jersey, since the Breamer burglary.

Next day, checking with the oil company, we learned that the card had been used to buy gas on the very night of the burglary—only a block from the Breamer apartment. Recorded on the credit card slip was a license plate number registered to a stolen car.

How could the Shooter be sure that the gas-station attendant, only a block from the burglarized apartment, didn't know Mr. Breamer personally? The Shooter didn't have to be sure. He didn't care. If the attendant challenged him, he could always pull out his .25.

The rest of the oil company record told us more of even greater interest. The card had been used twice, since the burglary, at a gas station on the Grand Concourse in the Bronx. The Bronx? Where the old man in the car was shot. The license plate number recorded on those slips was the same one used on the night of the Breamer burglary. He still had the gun. He still had the card. And he still had a stolen car.

That night a stakeout was set up at the Bronx gas station. The attendants were carefully instructed to signal to detectives parked nearby if anyone presented a credit card bearing Abe Breamer's name.

Two nights passed uneventfully. On the third, Detective Walter

Cicack approached a red Pontiac convertible in which a young man sat while his tank was being filled. As the cop reached the pumps, he identified himself and asked for the man's license and registration. The suspect leaped out, gun in hand. Cicack tried to grab it, they struggled, wrestling to the ground. The gun went off, a bullet striking Cicack's hand.

Another detective, running up, crouched behind a gas pump. The suspect shot at him, jumped into his Pontiac, and pulled out. He raced to the Major Deegan Expressway, and though the police gave chase, they soon lost him.

At the gas station scene, ejected shell casings were picked up. They were from a .25-caliber automatic.

The stakeout cops had fired more than twenty shots at the fleeing gunman's car and several seemed to have gone through the rear window. Hospitals and doctors throughout the city were alerted to be on the watch for anyone appearing for treatment of bullet wounds.

Next afternoon we had it.

From a scared stoolie in the Bronx. A young guy he knew had cut a bullet out of his own forearm. Not a police bullet. A bullet from his own gun. Said he'd shot himself in an accident.

The guy's name? Barry Schwartz.

Where did he live? The stoolie shrugged. But within an hour we had an address. Charlie, Frank, and I—along with at least ten other cops—waited for Barry to show up. After two days of the stakeout proved fruitless, I picked the locks on his apartment and went in. Here's what I found:

A crisply made bed
Fresh, clean curtains
A small hooked rug
An empty box of .25 automatic ammunition
Two Garrard turntables, one with the wires pulled out (loot)
A drawer full of complexion creams and skin ointments
Ladies' costume jewelry (loot)
A jar full of pennies
A ski mask
Newspaper clippings on the death of James Braddock
Empty drawers. And no suitcases.

Our quarry was gone.

It did not take us long to work up a profile of Barry Schwartz. He was twenty-two, with a history of mental illness. Disappeared from his usual haunts—a pizza joint and an ice cream parlor popular with girls in the neighborhood. No close friends. People he did see with some regularity were frightened of him. His parents lived in Flushing (not far from the Braddocks) and some years before they had rented a cottage in Fallsburg, New York, a small resort town in the Catskills.

We figured Fallsburg might be a good bet for a disturbed young man on the run. We teletyped Police Chief Seymour Farber there to watch for Schwartz—New York City's most wanted man—and his stolen car. We couldn't find the red Pontiac. Barry might still have it, bearing the same plates, just as he still must have the gun.

A few days later in Fallsburg a teenager phoned Chief Farber. Someone was flashing a lot of hundred-dollar bills at the Hillside Motel. Checking the license plate on the "someone's" car, Farber confirmed it was stolen, the one we had notified him was being used by the Shooter. Then, with a few other cops, he burst into the motel room, guns drawn.

Barry was seated on the toilet. The .25 was in his pocket. He had no chance to reach for it.

Charlie Prestia and Frank Moore rushed up to Fallsburg and started questioning our man. Later, Charlie described him: "Face of an old juvenile, eyes sly, but clean-cut, like a college football star dressed up for a date."

To our detectives Barry admitted to burglaries but at first refused to talk about any shootings. Yet Barry said of himself:

"I know I'm dangerous because I don't look like I'm vicious. People trust me and that makes me especially dangerous. If I had had any warning before Chief Farber broke in I sure would have shot it out."

Assistant District Attorney James C. Mosely of Queens drove up to take a statement in the event the detectives got a confession. The Shooter held out until a local lawyer was assigned to him and Barry was told what Mosely had in evidence against him.

Finally Barry agreed to a confession, adding with a grin, "Well, I can always swallow some glass."

From the questioning by the Assistant D.A.:

MOSELY: Did you shoot a man in the Bronx?

BARRY: Yes.

MOSELY: Why?

BARRY: I just felt bad. I felt like hurting someone so I went back and shot him.

MOSELY: What about James Braddock?

BARRY: I entered that apartment through a window. I didn't know it was vacant because the blinds were down and no lights were on. For a few minutes I was walking around the apartment. Then all of a sudden, you know, somebody came into the room, and I believe the man was taller than I was, and a little stockier. He came up on me suddenly and my heart started to pound. The adrenalin in my system, when, you know, somebody says boo to you . . . and I jumped. I almost automatically . . . without any thought . . . I just grabbed the gun. I fired. . . .

Barry Schwartz, burglar and shooter, was convicted of manslaughter and sentenced to ten to twenty years.

There is an epilogue to this story. Barry was serving his stretch at Attica state prison when the explosive riots erupted in 1971, during which forty-three men met violent death. State investigators gave me this account of what happened to him:

The day before the insurrection broke out, Barry and another inmate, Kenny Hess, were interviewed by Don Stuart, a Buffalo TV newsman. The interview, conducted in the Attica yard, dealt with a fight they had witnessed that day between two other inmates. In the midst of the interview, the newsman's notes were seized by Herbert X. Blyden, a black who had been a central figure in the inmate takeover of New York City's Tombs prison the year before.

Next day, when the Attica inmates rioted, Barry and Hess were dragged before a kangaroo court composed of some thirty prisoners who had appointed themselves the "Security Guard." Blyden accused the two men of being "traitors" for telling the TV newsman about the yard fight. Both men were stripped, marched naked into a cell, and beaten. Later that day they were paraded through all of "D Block," again beaten, and probably raped. Broken glass throughout the cell block cut deeply into Barry's bare feet. Outside medical aid was called in and Barry's feet were bandaged.

On the second and third days of the riots, Barry and Hess were kept in "custody" and beaten often. On the fourth and last day, the Security Guard decided it was time the two "traitors" died. Barry and Hess were thrown into a cell and their throats cut. Then each member of the Security Guard stabbed them, a symbolic action of unity—"all of us are in it." Barry died on the cell floor.

After the state police gained control and the riots ended, Blyden and three other inmates were indicted for the murders of Barry and Hess. The charges: kidnapping and felony murder. A felony murder is one in which a person commits a murder while in the act of perpetrating a felony. Early in 1975 a state supreme court justice ruled that the prosecution had failed to prove Barry and Hess had been kidnapped and that therefore a felony murder did not occur. On the basis of this judge's ruling, because the indictment called for "felony murder" with kidnapping specified as the felony, the charges against the accused killers of Schwartz and Hess were dismissed.

The Shooter, certainly, had not proved invincible.

What had driven Barry, the homicidal burglar? One clue may come from the TV interview he gave under volcanic circumstances. He seemed to have craved some kind of recognition as a "somebody," a theory supported by his self-portrayal as "dangerous because I don't look it." He must have been a dreamer and in his fantasies saw himself as superior to ordinary people around him.

Here again was a member of the boldest profession, an aggressive thief with no discernible sense of inferiority—a burglar with a death wish ultimately granted by his peers.

But what about his victim—and others who succumb to maniacal burglars like Schwartz and Robles?

The Wylie-Hoffert murders and the arbitrary killing of James Braddock show us two sides of the coin. In the first, there's little doubt that a different reaction by the young women—and most certainly by Emily—might have left them alive. But the Braddock crime—this wanton, illogical, senseless killing? Probably in no way would Braddock have escaped Schwartz unscathed, any more than you can prevent some deranged person from walking up to you and shooting you on the street, in a bus, or through your window. You can, however, do your best to keep intruders out of your home with all the anti-intrusion measures now available to you.

Friends, Romans, countrymen,
lend me your ears . . .

Chapter 8 When you're bugged or wiretapped

YOU'VE SEEN and read about eavesdropping bugs and telephone taps in the movies, on television, in spy and suspense thriller books. The White House under President Nixon conceded it had secretly authorized seventeen wiretaps, including the phones of five newsmen. The FBI and CIA have long used these measures. In the Pentagon Papers affair, when the government admitted that Daniel Ellsberg had been wiretapped by the FBI, the case against him collapsed. Police and other law enforcement agencies, presumably armed with court orders, bug and wiretap suspected criminals.

How can such invasions of privacy affect you?

Eavesdropping can be regarded as a form of burglary when a device penetrates a room, and certainly it can lead to burglary. Quite often, planting a bug begins with surreptitious entry, as in the Watergate fiasco. A couple of years ago the Rockefeller Commission heard testimony describing a break-in against a CIA employee who had attended a meeting of an organization suspected of having links to a foreign government. As the Commission reported:

"A *surreptitious entry* was made into the employee's apartment by cutting through the walls from an adjacent apartment so that microphones could be installed. Seven microphones were installed so that conversations could be overheard in every room. . . ."

Eavesdropping may intrude in your private life, too. Evidence for divorce proceedings has been obtained this way. Domestic troubles have spurred the use of listening devices. During the 1976 Presidential campaign, Alabama's governor George Wallace confirmed that a bugging device and some tapes had been found at the governor's mansion. "This happened," said Wallace, "in my bedroom between me and my wife. As long as it doesn't affect the State of Alabama or my services as governor, it remains the business of me and my wife."

According to news reports, when Mrs. Wallace learned that her jealous husband had placed her under "surveillance," she retaliated by rigging up a crude phone tap. Acting on a tip, Wallace ordered security officers to check out, or "sweep," the mansion. In the basement they discovered $2,000 worth of recording equipment hooked up to the governor's bedroom phone and a business phone. More than two hundred reels of tapes were also uncovered in a safe.

Amateur stuff, obviously. Far more sophisticated and certainly considerably more prevalent is the practice of eavesdropping in business and industry. Conceivably you can be a victim in your office or plant. Companies bug and tap competitors for industrial and trade secrets—for everything from new product developments to potential mergers, new processes or formulas, stock discussions, and plans for expansion. The American Management Association once estimated that industrial listening-in and business theft takes an annual toll of $2 billion. It is most common in the automobile industry, among clothing manufacturers, between advertising agencies. And industrial espionage is apt to involve trespass, breaking and entry, and theft.

There is, too, much prying among rival executives and of executives listening in on employees suspected of wrongdoing.

Not generally realized is that in most states it's perfectly legal to eavesdrop on a room conversation just as long as one of the participants is aware of the bug. The morality of such snooping is, of course, something else again. But apparently many executives feel that in their dog-eat-dog world the ends justify the means.

The bugs and where they're hidden

The basic components of electronic spy devices include:

A *transmitter*, which is a wireless instrument by which sound is picked up and transmitted by way of radio waves through the air.

A *receiver* picks up the sounds transmitted and converts them to sounds the listener understands.

Microphones pick up sounds and convert them to electrical impulses.

A *speaker* is the device from which we hear the sounds.

An *amplifier* amplifies or increases a signal.

Added to these are specialized equipment such as recorders, audio probes, telephone pickup coils or induction coils, parabolic mikes, or gun-barrel mikes.

Devices used are so sophisticated that a transmitter can be hidden in an ashtray or even in the olive in a drinking partner's Martini, the toothpick sticking up acting as an antenna. Espionage mikes, the size of a hearing-aid earpiece, can easily be concealed in the fabric behind the easy chair you're sitting in, and some come with sharp hooks for the purpose. A pinhole or keyhole mike known as "the snake" or a "spike mike," which minutely penetrates a wall, defies detection. For wall listening, an audio-stethoscope-type mike is popular. Perhaps the smallest transmitter is the size of a postage stamp. A "sugar cube" transmitter can pick up a whisper at twenty feet. (See illustration on page 162.) One room bug, the "Snooper," is said to be able to pick up the riffle of a newspaper and transmit it clearly for two hundred yards. And there's equipment that can pick up conversation off the surface of windows. A parabolic or gun-barrel mike is pointed at an open window to listen to gossip or your latest confidential plans.

Where are these contraptions installed in your office or business premises? Detective McDermott, once asked by a corporation president to "sweep" an executive board room, discovered a bug in a mounted swordfish—the fish's steel spine was used as an antenna. Mikes have been found in toilet-paper rollers in a company's washrooms; by throwing a switch, ten toilets could be monitored. Bugs have been used in draperies, under a table, behind wood paneling, in conference rooms, in inkstands and ballpoint pens. In larger offices, homes, and factories, favorite places are in air conditioning and heating ducts. Usually the bug is taped solidly to the inside of the duct, about ten inches from the opening.

To install listening instruments, a favorite ruse of industrial spies is to pose as a repairman, appropriately dressed, sent to fix an air conditioner, furnace, or phone. When he has to resort to surreptitious entry, and must return to renew a transmitter's batteries, the intruder is likely to remove the cylinder of a door lock and make keys for it. Then he avoids having to pick the lock again.

Radio transmitter mikes can be installed in seconds. The wireman has only to drop in on your premises, reach behind a drape, and hang the bug in its folds. The trouble with these mikes is that the batteries need periodic replacement. Thus a long-term surveillance may well call for a hard-wire carbon mike installation with the mechanic having to run wire to a telephone block in your office or home. When this is necessary, installation takes quite a bit of time; rarely will a "repairman" get away with that long a visit. Here, then, is where the pick man comes in and—if there is no court order—burglary is committed.

The last thing a wiretapper or bug-installer wants is for you to know he has been in your place by forcing his way in; such an

Equipment used to install eavesdropping devices: **Center**, two "sugar cube" mikes that will broadcast by radio frequency for several blocks; **Left**, telephone head set to identify victim's pair of telephone wires so that a "wired" carbon mike (upper right) can be installed; **Bottom**, flat steel tool is "shove knife" to open spring latch.

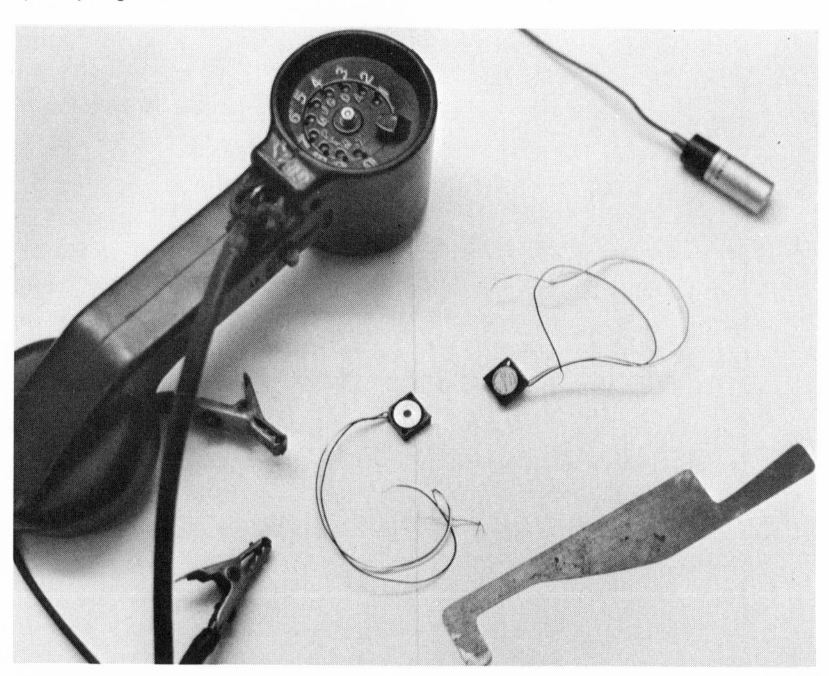

obvious invasion may make you wary of a tap or bug. When McDermott uncovered that wireless transmitter in the mounted swordfish he also found that the locks on the door of the boardroom had been picked. Executives who believe that their offices are sacrosanct because a guard is on duty in the building's lobby all night would do well to pick up some lessons from McDermott's experiences.

If Mac had to get into a guarded building he would signal the watchman to come to the front door, show him telephone company "credentials," and say he had to go to the basement. Dressed as a repairman, Mac was never challenged. The watchman had to be at his station. Once in the basement Mac opened up the phone company's equipment boxes, strewed some tools and wire on the floor, and made his way up the stairs to the office he had to break into.

If a job came up suddenly and Mac couldn't "uniform" himself in time, he resorted to another ploy. Picking the front-door lock of the office building, he approached the guard as if he had used a key. Announcing he was going to the "Jones and Smith" office on the sixth floor (a real company and its real floor), Mac would sign the visitors' book "James Valentine" (or any name for which he was carrying a business card). Then he would take the elevator to the sixth floor, pick open the "Jones and Smith" office door, go in, put on some lights, and leave, locking the door behind him. Mac would walk up to the seventh floor for his real target. After completing his assignment he returned to the "Jones and Smith" office, put out the lights, and departed.

How to tell if you're being bugged

Chances are, you'd start by looking around desks and picture frames, but they're too obvious for plants. Thoroughly search the bases of lamps, bottoms of wastebaskets and tables, and in ventilating ducts. Notice especially if anything has been recently added, such as a new phone. Look under carpeting and rugs. Take all drawers out of your desk. Examine recent work areas for new paint or plaster on walls and moldings. Look for any tiny hole in a wall, probe into it, and get a look at what's "next door."

If a repairman has been around, check the area he has visited.

Inspect any loose wires or wiring for bugs such as the hard-wire carbon mikes. (It would be a very thin wire.)

So you still can't spot signs of a bug? Start at one place in a room and probe every square foot, alert for any irregularities, flaking, or too-clean spots. Go over your windows and doors, to see if you can detect putty marks, nail holes, or signs that a piece of molding has been removed and replaced. Plates from electric outlets should be removed and the boxes inspected.

If nothing shows up and you are still suspicious and determined, you can go so far as to obtain eavesdropping-surveillance equipment; a kit can be purchased from Saber Laboratories (1150 Bryant Street, San Francisco, California 94103) or Dektor Counter-Intelligence, Washington, D.C. If you learn how to use the equipment you'll be able to detect the presence of transmitters. Your alternative is to call in a team of professional countermeasure experts.

We strongly suggest that you don't just get the name of a company performing detector services from the yellow pages. Some services, preying on your fears, are virtually useless. People who have been engaged in law enforcement are preferable, particularly those former officers or agents who have been honorably retired or otherwise separated. A good criterion is membership in accredited professional organizations, such as the Professional Investigators Association and the American Society for Industry Security, by principals of the countermeasure company. Determine from the prospective service if it has been regularly employed by any particular reputable company, and ask for references at the police department in your community. Usually they'll cooperate.

Remember, there is no magic wand—no piece of electronic equipment that, when introduced into a room, will detect any and all bugs. A good "sweep" requires a physical search as well.

"My phone is being tapped!"

Every year U.S. telephone companies receive about 10,000 complaints from subscribers who believe their phones are being tapped. The practice has been commonplace in Washington, in Moscow, in embassies throughout the world. As with bugging,

Telephone with pocket tape recorder.

phone surveillance is used in marital conflicts, among business rivals, and in law enforcement. A suspicious—or paranoid—company executive may tap employees' phones. More than one corporate president has admitted recording every phone call made to him—and it's legal, provided that he is the only user of that telephone (see photo).

So-called in-line telephone transmitters can be attached to a phone either across the wires or on the terminals in a junction hookup, popular in apartment and commercial buildings.

A tiny transmitter, an exact duplicate of the carbon mike in every phone, may be substituted in ten seconds for the regular unit. Just unscrew the mouthpiece, drop out the telephone company version, and put in your own. Both ends of telephone conversations will then be broadcast for several hundred yards.

With the amazing "harmonica bug," an eavesdropper can dial your bugged phone from as far off as a thousand miles away, or any distance, provided there's a direct dialing system. This is not a device for tapping the phone itself; it permits snooping on conversations carried on within about four feet of the phone instrument. Once the phone instrument is bugged internally, simply dialing its number and sounding certain notes into the phone will enable a listener to hear everything said near the phone. Blowing a particular note on a key-of-C harmonica activates the phone after dialing—and no ring from the telephone will alert the victim. The eavesdropper becomes part of the room conversation until the phone is used. Lifting the receiver to make an outgoing call disconnects the unit. After you hang up, the invader merely has to dial your number and sound his note to be back with you again.

While a tap of a telephone is most often made by someone disguised as a phone company repairman, electrician, or building inspector, surreptitious entry is not at all unusual. But with an in-line transmitter the tapper doesn't have to break into the premises at all, or even tamper with the phone instrument itself. Using radio waves to transmit sound to a receiver located elsewhere, the transistorized unit can be installed anywhere along the line or in the telephone junction box. All that's necessary for the business or home phone to be tapped is that the proper telephone line be identified.

Devices are made, designed to overhear both sides of your phone conversations, which require no wiring at all and are in no way physically connected into the circuit. They only have to be placed within three or four feet of the instrument. Properly disguised or hidden, they are difficult to detect. Here, again, the tapper has to get into your office or home.

The wiretapper who got tapped

Criminals themselves—particularly the organized element—have long employed wiretaps and bugs for their operations. For decades a thorn in the side of New York police was a character known as "Cheesebox" McCarthy. Short, slight of build, with the nervous energy of a rodent and the slyness of a fox, Tim McCarthy was the underworld's expert on bugging and wiretapping.

McCarthy's strong suit was the telephone. His services were constantly in demand by gambling figures throughout the country. He could guarantee that the phones in wire rooms (gambling action centers) could not be traced if one of his "cheeseboxes" were used. Here's how his gimmick worked:

A bookie would rent a small office anywhere in the city and give McCarthy the address. Visiting the location, McCarthy would plant one of his devices on the telephone line coming into the vacant office. Then he'd put an alarm trip on the door, wire it into his installed "cheesebox," and his task was done. The gamblers were ready to start business. The phone number of the vacant office was given out to the bookie's customers as the number to call to place a bet.

Early each morning the wire-room employees dialed the same number and then merely left their phone off the hook. All day long any calls made by horse-players to the vacant office would automatically be received at the gambler's location. When the number used by the customer became known to the police it was traced and the given address raided. Now the bookie's investment in McCarthy really paid off. Police coming through the door tripped the alarm, and a signal went through the phone lines warning the wire room there was trouble. The bookmakers would then simply hang up their phones; all the police would have was Tim McCarthy's "cheesebox."

Investigators figured that one way to get McCarthy might be through a wiretap on his own phone. Chances were, he was so confident of his ability to detect taps that he would be unguarded in his conversations. In fact, taps on his home phones were tried, but McCarthy was aware of them almost before the police experts had completed their installation. Then came a break. McCarthy was planning to move to a new apartment. Detectives, finding out the address, asked the telephone company to notify them when the new phone would be installed. On the day of the installation, while McCarthy was still supervising the moving men, the police obtained a court-ordered tap on his new phone.

That night, detectives monitoring McCarthy's tap from a building a block away noticed some strange things. The tape recorder connected into the line would start to roll, stop, then start again. This went on for more than an hour. At three o'clock in the morning one of McCarthy's sons made a call into the apartment from a public booth just down the street from their building. McCarthy answered:

McCARTHY: Hello. That you?
McCARTHY, JR.: Yeah. How's it sound?
McCARTHY: Don't talk. Just hang up and call me right back.

The McCarthy scion did as he was told and his second call was answered on the first ring:

McCARTHY: O.K. Did you get your dime back?
McCARTHY, JR.: Yes.
McCARTHY: Good. That's it. I'm all set, you can come home.

What McCarthy had done was to use instruments to measure impedance, resistance, and the other electrical values on the new telephone line, to establish the norm. Then he hooked up his detection equipment based on these readings. As long as these values did not change (as they would if anything was thereafter introduced into the circuit, such as a wiretap) he could be sure his line was "clean." But all of his readings were made with the police tap already in the line, so the tap became McCarthy's "normal" condition—and he did not detect the tap. The return of the caller's dime was an additional check on his detection circuitry.

From that day on, when McCarthy received an incoming call he had the caller hang up immediately and call back again. His question to the caller was always the same: "Did you get your dime back?" When the puzzled caller said he did, McCarthy said, "O.K. My phone is O.K. Now I know the one you're on is clean, too. So talk to me."

Business for McCarthy was apparently slow at that time so it took the police about a month to nab him. He was grabbed in a vacant store just finishing up a "cheesebox" job. He couldn't believe the police had gotten him on a wiretap. In the courtroom McCarthy changed his plea from not guilty to guilty as soon as he heard his recorded voice. He took his medicine.

Recently Tim McCarthy wrote a book about his exploits. He neglected to mention how the wiretapper got tapped.

McCarthy and his breed have long been with us. Watergate only made the public more aware, just as the arrest of a fellow named Steve Broady made New Yorkers nervous years ago when he had the temerity to tap the phones at City Hall.

Detecting a tap

Most telephone taps are easier to discover than phones used to bug room conversations. If you complain to your phone company they will make an honest but rather perfunctory check. They uncover only about two hundred taps a year. When an illegal tap is found, you will be informed; some phone companies may tell you if the tap is legal—done with a court order. Nevertheless, inspectors rarely do a thorough job in checking the inside of your phone and any sophisticated device goes undetected. For instance, by transposing a couple of wires your telephone instrument becomes a permanent room-bug. This can be discovered only by opening the instrument.

Most taps are direct wire jobs—wires directly connected to your phone terminals. A phone tap can be located by physically checking for foreign wire connections on the line or measuring line impedance or resistance, similar to "Cheesebox" McCarthy's methods of detection. One firm puts out a device called the Teletale, a phone line indicator that shows whether or not the line is "in use" when your receiver is still on the hook.

Also available is a meter readout indicator, The Private Sentry, which needn't be connected directly to the phone lines. You merely put the probe in contact with any metal surface on the bottom of the phone. The meter needle will deflect if your phone is being tapped.

If you know, or learn, what the components inside a phone should look like, you can remove the base of the phone for inspection to see if something's different. If you think a tap has been placed elsewhere, check the terminal box on the wall outside or in the basement of your home. In an apartment, find out the number of your "house pair" and check the basement box. If you notice more than one wire in each terminal, you have some cause for concern.

PRECAUTIONS TO TAKE

► To frustrate a sophisticated phone tapper, you might consider a speech scrambler. This is a portable, fully transistorized device that electronically scrambles the spoken voice into wholly unintelligible sound patterns. It's held against the phone and changes your voice into sounds that can be unscrambled only by a matching device on the other end. Only you or people you trust who have an identically coded scrambler can get a clear voice message.

► Never have a phone in a boardroom and don't allow those attending meetings to bring in attaché cases. In companies where privacy of conversation is imperative, the room should be "swept" and physically searched. Maintain strict key control, use high-security cylinders to actuate deadlocks, and keep the room locked when it is not in use. Have a responsible employee present when the room is opened for cleaners. Available are prefabricated isolation booths designed to preclude eavesdropping; these small "rooms" can seat four comfortably.

► Borrow or rent a small mine detector at a supply house serving builders. Anti-intrusion experts use them to note the change in the pitch of sound if the instrument is passed close to a transmitter. However, it will not detect direct wired mikes.

► To drown your voice for a bug, resort to the familiar tricks of turning on your shower, playing a nearby radio loudly, or simply rustling cellophane in your hands.

► Buy a small neon sign and turn it on near you when you're on the phone. The listener-in will hear only the loud buzzing radiated by the sign's transformer. (Note: This works only with an induction coil, not with a wired tap.)

Let's say you're still not satisfied and you insist on removing that wiretap. Here, too, try retaining an electronic countermeasures firm. Be sure you get one that's really competent because rip-off charlatans have infiltrated this field.

WHEN TO CHECK WITH YOUR PHONE COMPANY

► Someone claiming to be a repairman comes to your door. You have made no service complaint. Do not rely on his phone company I.D.

► Your phone rings and a voice says, "I'm just checking," and hangs up.

► You notice a man working near your wires or on a telephone pole and he is using a private car. (Better still, for this fellow, call the police.)

► Telephones are rarely tapped casually and usually are skillfully installed. Unfortunately, many people are almost paranoid about eavesdropping and wiretapping. You can worry most when your phone reception is uncommonly good. But if you have an independent reason to think your phone may be tapped because of specific business or personal situations, call your company if you hear *persistent* "funny noises" on the phone. Such odd sounds may occur in stormy weather but they should not continue and may be the result of a *crude* wiretap.

► Other tip-offs to a tap: On inept taps there may be unusual delays, strange scratchings, or an echo. Faint intruding voices you hear sometimes are due to an imbalance in the circuit caused by unskilled wiretapping or to faulty monitoring equipment. (You can also notify the FCC.)

Operation fortress: bugging a stronghold

To a highly skilled break-in artist, planting a bug is generally a cinch. Even when the target is a well-fortified home or office it may not be unimpregnable. Detective McDermott, designated by the New York City Police Commissioner as the department's master burglary expert, was frequently called upon to make a surreptitious entry in order to gather evidence against suspected criminals. Occasionally he was directed to open doors so that his colleagues might more safely approach and arrest criminals disinclined to be taken alive.

In bugging and wiretapping cases, he was granted a court order authorizing him to "make connections with any and all wires leading to and from the aforementioned premises and to do all things necessary to permit conversations in the premises or the communications being transmitted over the aforementioned telephone instruments to be intercepted. . . ." The wording of the court order and search warrants made necessary the picking of locks, the opening of safes and file cabinets, and the compromise or circumvention of alarm systems, combination locks, and the like. During his long career, McDermott estimates that he made about a thousand such surreptitious entries.

On one occasion McDermott was called upon to contrive entry and to help install eavesdropping devices in an apparently burglarproof apartment, a stronghold of the Mafia. Here is his story:

One morning Benny, the wireman (electronics expert) of a police command, called me.

"I've a tough one for you, Mac," he announced.

That was sheer understatement, I was soon to learn. The target assigned to me was a midtown penthouse maintained by a notorious *caporegime* for meetings with out-of-town contacts. This Mafioso, whom I'll call "Don Cheech," was usually surrounded by "muscle" and the pad was so secret that only the capo's most trusted lieutenants knew of its existence.

Benny the wireman, who invariably dressed as a telephone repairman, lived his work and was constantly experimenting with new ways to bug and tap, confining his research to the kitchen table at home. He never wasted any time on a job. A veteran of fifteen years on the force, Benny had worked as a telephone man before joining it. Modestly, he often said, "I knew nothing until I got into wiretapping in the Police Department." A cigarette addict, it was hard to keep him from smoking on a job; a cigarette in a toilet bowl or even smoke in the air could be a giveaway.

At dinner that evening Benny told me more about the assignment.

"This penthouse apartment," he explained, "has been used for only two previous big meetings. We believe the topic of this upcoming meet is the setting up of a new smuggling network for heroin. We've got solid information that the route is to be from Marseilles through Paris to Montreal and then to New York, with parts of the load going from New York to Detroit and Buffalo. Our purpose, Mac, is to identify the courier and the method of smuggling.

"The reason this penthouse pad was chosen for the negotiations," Benny went on, "is the absolute security of the building. Now, there's a little complication for us. This guy Cheech is very friendly with a woman who occupies the adjoining penthouse. And she has a habit of walking in to see him unannounced."

I asked, "What exactly is our goal?"

"We've got to plant mikes in Cheech's place. We already have a court order for the wiretaps and we'll get a bug order before we go in."

From that night on, I was briefed daily on the building's formidable security and on Cheech's movements. A single entrance to the building was diligently guarded by a hefty "receptionist" whose job was to interrogate every person entering whom he did not know. Each visitor was required to state his or her destination; the guard would contact the tenant via the house intercom, to verify the visitor's identity. Only then could the caller be allowed to use the elevator. The guard and his night counterpart never left their post to hail cabs for tenants.

No service entrance. No basement garage. No fire escape. No possible access from adjacent buildings. We nicknamed that building "the Fortress."

Surveillance of Cheech soon determined that he rarely used the penthouse except for conferences.

One of the detectives assigned to watch the building so that I could be kept up to date was handsome Tommy O, a tall Irishman. A bachelor, Tommy found himself distracted from his duties by the emergence from the Cheech building each morning of a willowy blonde. The girl became an obsession with him and a topic of conversation among others on our task force. Finally, it was decided that such a splendid model of womanhood deserved to meet one of New York's Finest. Tommy, single and smitten, should let his existence be known to her. Maybe this could be worked up; maybe this was the "in" we were looking for.

Tommy O accepted the assignment of Trojan Horse with alacrity. The next morning he followed his subject to her office on lower Broadway and for several days watched for a chance to contrive a "natural" encounter. He noted that she lunched daily at the same restaurant but always with another woman. On the third day, she lunched alone, at a table for two, in the crowded restaurant. Tommy O casually walked over and took the vacant seat. By the time he had his second cup of coffee, the smiling

Irishman had used his blarney to strike up a warm conversation with his subject, Sheila.

Tommy, she was informed, "worked on Wall Street," had lunch at the same time as she did every day, and lived but six short blocks from her own apartment building.

That Saturday night all the detectives agreed they made a striking couple when Tommy escorted her home from a movie. A few days later, getting out of a cab in front of her building, Tommy's jacket got hung up on his gun butt and it seemed the jig was up. Over coffee in her apartment Tommy decided to confess that he was a police detective (though refraining from revealing his assignment). He had not told her this before, he explained, because a girl he had previously dated had hated his "dangerous job" and worried about him.

Sheila was delighted. "Your job must be so fascinating," she told him. Tommy picked up the lead. Wouldn't she like to meet some of his fellow workers? His own apartment was so small, perhaps she could be hostess to three or four detectives and their "women-folk" (his word) for a little party. Very soon. Sheila agreed. A date was set for the party.

Meanwhile we were working, anxious to move. Two weeks before, a wiretap had been installed from outside the penthouse at a place where Cheech's line bridged in a terminal box three blocks away. It had proved unproductive: Don Cheech was being seen using public telephones all over the city. When he used his own phone it was only for innocuous conversation. It was imperative to plant a microphone in his apartment, and if it took much longer to do the job, the planning for the big new heroin operation might be completed before we could gather any evidence.

On the following Friday night detectives equipped with walkie-talkies had Cheech under surveillance at a fashionable restaurant, the Four Seasons, where he was having a leisurely dinner with a male companion. A police lookout car with similar communication equipment was parked on the street near the Cheech building. A second team was stationed on the roof of a building in the next street. From this latter point the windows of the penthouse could be observed.

At 8:30 the detectives watching the front of the building,

saw me, Benny, and three other detectives arrive, accompanied by four attractive young policewomen, nattily dressed. Tommy O had arrived earlier. Days before, he had told Sheila that one of his colleagues insisted on bringing the hors d'oeuvres. The policewomen had gone all out and were now laden with goodies. Under the hors d'oeuvres were the wiremen's tools and the equipment to be installed; my tools were in my shirt pocket. The building's husky receptionist, assiduously performing his duty, challenged us. Sheila, having previously advised him of the expected arrival of her guests, authorized our ascent. We were in.

There were the customary introductions, then drinks. Tommy O ruefully announced that he had neglected to buy dry vermouth and would have to go out for it. Benny and two other detectives volunteered for this chore, saying that Tommy could make Manhattans in the meantime. I expressed concern for my car, parked in a restricted area, and said I'd go with them to move it. Leaving the apartment with the tools and a walkie-talkie under our coats, we made our way to a fire stairway and climbed several flights to the penthouse level.

We knew that Don Cheech was out at the restaurant but his next-door lady friend had not been seen all day. No sound was coming from her apartment and the lookouts had earlier reported that it was dark, except for an exterior patio light. Cheech's door was fitted with two locks, both with conventional cylinders. Prominently displayed on the surface of the door was a decal informing us that the security-conscious Cheech had an alarm system. Were all our efforts then for nothing?

Oh, no. This, I recognized, was a local alarm that would sound an audible device upon intrusion. But the Don could use a "special key" in a shunt switch (equally obvious) mounted on the frame of the door. With it he could shut off the device before entering, to avoid the bells disturbing his neighbor.

I approached the door while Benny and the others listened for an ascending elevator, covered the fire stairway, and monitored our walkie-talkie unit. Using my tools I picked the key-controlled alarm switch to the conveniently marked "off" position and then picked the two door locks. After I slipped a small piece of metal (the end of a paper clip) into the keyway of the upper cylinder, we entered. I locked the door from within.

Forty-five minutes later we were still in the apartment. We needed a hard-wire bug, not a battery-operated transmitter that would require us to get back into the place about every five days to change the batteries. Running wire along baseboards and under rugs, wherever we tried to place the mike, the reception was poor. We had the lookouts on the street and rooftop giving us periodic checks so that we'd be assured our walkie-talkies were working, in case Don Cheech came into the building. Usually a hard-wire carbon mike, though harder to install, gave very satisfactory performance but in one place the radio was too close and in another the mike was too near Cheech's television set. While the bedroom was ideal it was necessary for the bugs to be set in the living room, where we felt anything of interest would be discussed by Cheech and his lieutenants.

Suddenly our walkie-talkie came to life. Our equipment was strewn about the living room and the lookout on the roof in the next street was telling us, calmly, "Cheech's girl friend next door has come home."

I told the technicians to keep working, acknowledged the call, and took up a position just inside Cheech's door. About five minutes after her arrival home, Cheech's woman elected to visit him. She knocked discreetly. I was certain she must have heard us moving about and had concluded that he was home. The other detectives didn't hear her knock. Looking out through the one-way peephole I was confronted by my first sight of the Don's "little friend," a big-nosed, olive-skinned woman in her thirties. Since she was standing close to the door I could not communicate to the others what was happening. Why didn't she ring the damned bell? They were working quietly enough but the walkie-talkie volume was still turned up. At this moment our street lookout decided to inquire of the roof lookout if "that bitch's lights are still on."

Peering through the peephole I could tell that the woman had heard something. She actually cocked her head to one side, as people will when they don't know they are being observed. Disappearing from my view she probably stooped to press her ear against the door. It was past time to let the others know that things were getting sweaty.

I hissed a warning to Benny, who was working nearest to me.

Our neighbor put her key in the lower lock. Now Benny and the others cocked their heads; we could hear the dead bolt being withdrawn. By now we had been in the apartment well over an hour.

Reacting automatically and suddenly, Benny reached down to pull his gun out of its leg holster. But he lost his balance and hit the wall, with a practically inaudible thump that sounded to us like the crash of doom. Fortunately, Cheech's girl was so busy rattling keys that she didn't hear it. Howie, one of the detectives, who has a peculiar sense of humor, ran laughing into Cheech's bedroom. (He later said he had to get away or break up completely but I think he just wanted to remove himself from Benny's line of fire.)

Our lady neighbor was now working on the last barrier, the upper lock. Thank God, I had inserted the end of a paper clip in that lock. Her key moved it to the back of the cylinder where it lodged against the tailpiece; this prevented the key from going all the way in. After another minute of alternately trying the key and listening at the door, she went back to her hutch and we were left in peace.

We quickly discussed our situation, particularly the odds against her calling the house management. Surely she wouldn't expose herself by complaining about Don Cheech's lock malfunction. I felt, too, that she would probably lack a key to the exterior alarm switch and knew from the position of the switch that the alarm was off—or she wouldn't have tried to open the door. She would almost certainly then believe that Cheech was home. But why didn't he answer her knock? Benny came up with the most probable course of her reasoning: "She now thinks Cheech has another broad in here, that he won't open the door and has locked her out." Agreeing, reluctant to abort after so much effort, we kept working.

Our wireman finally planted the bug effectively behind a bar. Dozens of bottles and glassware had to be taken down from the shelves and the entire wooden structure moved away from the wall and then replaced, concealing the mike. After confirming the quality of reception from the device, we signaled the lookouts that we were leaving the apartment. The others waited in the corridor for me to lock up. Benny placed his hand over the lady neighbor's peephole while I reset the alarm and actuated the two dead bolts by picking, after first removing the sliver of metal from the upper cylinder with a magnet.

Using the fire stairway again we arrived at Sheila's floor. From behind a fire hose on the landing we retrieved the bottle of dry vermouth placed there by Tommy O much earlier in the evening. Then, almost two hours after our departure, we rang Sheila's bell.

Naturally, and not unexpectedly, the reception accorded us by our female companions was extremely cool. Benny headed immediately for the bathroom where he used his walkie-talkie to report our safe return. He was told that Don Cheech had just returned—a close call. The police listening post (for the bug located in the same building as the tap equipment) heard him burping from his repast.

At Sheila's we then had one of the best parties I've ever attended, except, of course, anywhere my wife was along. Tommy O told us that after we had been gone for an hour Sheila hoped his friends hadn't "run into something criminal." Very unprofessionally we decided to confess to her that we were with her under false colors, revealing the roles we were playing. Sheila was quite upset at first, then took it like a trooper.

At 3:30 A.M. we phoned the detective inspector to report our success. He asked to speak with Sheila and thanked her for her "cooperation." We could all relax now and discuss the night's accomplishments. Sheila, in the midst of the accolades we were bestowing on each other, told us she was glad it had been "business" that kept us away so long. "I wouldn't like to think," she said, "that Tommy's friends would be guilty of rudeness."

Ten days after our escapade Don Cheech decided to have his apartment painted. The detectives monitoring the bug could hear the workmen arguing over whether or not to move the bar away from the wall in order to paint behind it. The more ambitious painter prevailed and they came upon our hard-earned microphone—solidly taped to the wall. Using the tapped telephone, they called Cheech and he came on the run. The bug went dead. The last thing we heard on the tap was the Mafioso calling his lawyer to ask if it would be legal to use a shotgun on anyone he ever caught in his pad.

(Cheech's lady friend apparently never mentioned her trouble getting into his apartment. Almost certainly she believed he was in there with another woman and thought she'd leave well enough alone. You don't castigate a *caporegime.*)

The tangible results of this eavesdropping caper? Overheard conversations at Don Cheech's retreat enabled us to stuff the new pipeline, curtailing the flow of heroin. Two arrests in the United States wounded those in the Mafia engaged in international narcotic traffic, and the French police got a smuggler. Later, the Federal Bureau of Narcotics put Don Cheech away for a long fifteen-year stretch. Desperate to reestablish his heroin supply, he had become careless.

The rather elaborate planning essential for success in an eavesdropping Fortress Operation was the exception rather than the rule. A professional burglar will rarely take such risks even when the score to be stolen is considerable. The physical danger inherent in the kind of work we were doing is obvious. But, of course, we did not have to worry about being arrested, the prime concern of the thief.

For his security, and to keep out of prison, Cheech had depended on the limited entrances to his building, the "receptionists" on guard, the secrecy of the pad's existence—and on the completely inadequate lock cylinders and alarm system he had installed.

Chapter 9 "Official" burglary

THEY HAVE BEEN LABELED "government burglaries." To J. Edgar Hoover they were "black-bag jobs." Richard M. Nixon held that the President has the right to authorize a burglary anytime and anywhere on the basis of national security or "in the national interest." Generally, surreptitious entries are made either to plant bugs or to steal, photograph, or copy something like documents or files.

No doubt the most celebrated and most highly publicized official rip-off was the bungled Watergate caper. Five men from the so-called Plumbers unit got through a door at the Democratic National Committee headquarters, slipped inside, and began rifling the files, stuffing documents in boxes, dumping others on the floor. An additional objective, apparently, was to replace a malfunctioning bug that had been installed previously on the phones of chairman Lawrence O'Brien and another committee official. The Plumbers wore surgical gloves and carried walkie-talkies, two cameras, and electronic eavesdropping devices.

After using a piece of celluloid strip on a snap lock in a basement door leading from the garage into the Watergate office building, they rendered the spring latch inoperative by taping it to stay open, presumably so that they could come and go quickly. When a roving security guard discovered the tape he removed it, slamming the door locked. When he returned later, to his surprise the door was unlocked and had been taped again. Naturally the guard summoned police. When the cops, with guns drawn, burst in on the intruders, one of the five shouted, "Don't shoot, you've got us!"

Four of the "official" burglars turned out to have been either agents or operators for the CIA. The fifth was a simple Miami locksmith recruited for the mission. In a Watergate Hotel room from which the operation had been monitored, police confiscated a kit full of burglary tools, two pairs of work overalls, a wig, and a radio transmitter. Later, Lawrence O'Brien was to comment: "There is developing a clear line to the White House." The rest is sordid history.

Unlike the break-in of Daniel Ellsberg's Beverly Hills psychiatrist's office (in the Pentagon Papers affair), no lookouts had been placed to warn of any police confrontation.

The most incredible part of the Watergate burglary was its crude, amateurish performance. Newspaper accounts later told of large sums of money shelled out to finance the Plumbers' mission. McDermott's reaction to the conspirators' arrest was that they must have done so many of these jobs that they had lost their sense of reality. Apparently these men had been so dulled by repetitious performance that they had turned unbelievably careless. But after more information came out, Mac changed his mind.

These fellows, he concluded, weren't burglars as he knew them. They were dilettantes, virtually without talent for what they were doing. With almost limitless funds at their disposal, they recruited an ordinary Miami locksmith. The lock they taped open was a common one. It was taped so that their locksmith wouldn't have to 'loid it repeatedly (a matter of three or four seconds) as his fellow intruders came and went. To Mac this meant that none of the others were able to use celluloid on a simple lock. To have made this lock inoperative in such an obvious way, and to have done it on a door far removed from where they were working, constituted the height of stupidity. They left a break at their backs—in a building with a touring watchman.

Recent congressional investigations have revealed a considerable number of official burglaries by both the FBI and CIA. At a press conference, FBI director Clarence Kelley admitted: "Yes, the FBI has conducted surreptitious entries relative to the security of the nation." Testimony by Attorney General Edward Levi disclosed that the FBI burglarized offices of the Socialist Workers Party at least ninety-two times.

David Wise in his revelatory book, *The American Police State*, tells of an FBI man who said he was trained in surreptitious entry in an attic classroom in the Justice Department in Washington. There, FBI agents were instructed on how to make their own lock-picking tools.

According to a Justice Department task force investigating

the murder of Martin Luther King, Jr., FBI agents carried out five illegal burglaries to obtain information about him.

Investigating charges of alleged CIA domestic spying and unauthorized entries into offices and homes of U.S. citizens, the Rockefeller Commission heard evidence that the CIA conducted break-ins as part of training its agents. On one occasion, Wise relates, a CIA lock picker—without a search warrant—had difficulty getting into a Virginia photographic studio to take pictures of certain documents there. When he couldn't pick the lock, he removed the door's hinge pins.

Many mysterious burglaries, presumably "official," have occurred around Washington. Judging from their crudeness, these *known* burglaries may well have been the work of overpaid bumbling Plumbers. The home of Senator Howard H. Baker was invaded but no valuables were missing. In Chevy Chase, Maryland, an intruder rummaged through Senator Charles Mathias's personal papers. When the Georgetown home of CBS White House correspondent Dan Rather was broken into, his phone line was cut and someone removed two file folders from a cabinet. The burglar had carefully stripped the molding from a pane of glass in Rather's rear door, removed the glass, reached in, and unlocked the door.

In all these cases the intruders were committing burglary, officially or not. According to law, breaking and entering premises constitutes burglary when "the intent is to commit a crime therein."

Thus a police agent who enters a place by picking locks or using any other means of entry—and who doesn't have a court order for the entry—can be charged with only a trespass if his action was to further a police investigation and he had no intention of committing a further criminal act on the premises. If he achieves his entry without a court order for the purpose of planting an eavesdropping device, and the bug is also illegal, his action—whether or not in the furtherance of an investigation—would stack up as burglary.

There have been many instances, however, when the intent was to enforce the law, as FBI director Clarence Kelley implied. And this is where, for more than fifteen years, Detective Robert

McDermott performed multiple roles, ranging from testing the effectiveness of physical security at a missile site in the Southwest to his regular work where, on behalf of the New York Police Department, he made many documented surreptitious entries. In addition he plied his craft for the FBI, Secret Service, the Federal Bureau of Narcotics, the State Department, and other federal agencies.

During one afternoon and night he made four "penetrations" for different federal agencies and completed his tour at 4:30 A.M. on a job for detectives from a D.A.'s office in Greenburgh, New York.

None of these agencies knew of his work with the others. In the Police Department only his commanding officer and the Chief of Detectives were aware of these missions. Mac admits to occasional confusion as to "who belonged to what."

Agents of the Federal Bureau of Narcotics considered Mac's presence on a job as "lucky." He enjoyed working with them.

"You saw," he says, "an almost immediate result because they never witch-hunted. They were like the guys in our own S.I.U. [Special Investigations Unit, New York Narcotics Squad]—very sure of what and who they were after."

Mac relates having had only one bad time with the federal "narcs."

"They took me far out to sea on a Coast Guard cutter and I had to jump from the cutter to an open hatch in the side of an ocean liner. I spent the night endlessly opening passenger luggage in the ship's hold—and we came up empty-handed!"

One particular case presented especially sticky problems. Let McDermott describe his experience.

The case of the devious diplomats

When I was doing a job as an "official burglar," now and then I'd recall my encounter with Louie, a longtime savvy break-in artist, in a hotel lobby. As we chatted amiably he remarked: "Mac, the only thing I ever did really wrong was to have been born poor." Then he went on to talk about some of his reactions while practicing his trade.

"It's such a tremendous feeling when the job goes well—when the lock bolt slides back and you step into that room. . . ."

Louie and I had crossed swords in the past. Although I gave him his only stint in prison we were on the same wavelength. I had just come from doing a job in that hotel and was still full of the feeling Louie described. What troubled me at that moment was that Louie knew he was talking to a kindred soul. Two years before, when I had him on trial for a pick job, Louie introduced me to his wife outside the courtroom.

"This is McDermott—the other side call him 'Bobby Sticks' because he's pretty good with picks, too."

All this was said with a laugh but it was obvious Louie knew more about my work than he should.

Louie's young wife reached inside his shirt and fished out a religious medal.

"Look," she said, "we'll be O.K."

I stuck a hand under my shirt and showed her an identical disk depicting a broad-chested man tied to a cross. The legend on Louie's medal and mine read: "St. Dismas, Good Thief, Pray for Us."

The feeling Louie spoke of is insidious. I'm sure it's peculiar to pick men. You stand alone in that hallway or at that front door and concentrate all your energies in your fingers. The pulse is banging away in your throat. Your closed eyes, as you probe the lock, make it seem like you're in a trance. But, ah—when it's done and done quickly, and that final bolt goes back and the door swings open! You step inside and close the door, listening, then take a few deep breaths and the feeling is there. The "Duffy Sex Factor" is how I think of it—voluptuous and, yes, sensual. You move away from the door and penetrate deeper into the citadel. And you move from nervous fear to believing you're omnipotent.

I thought these ever present emotions couldn't be further complicated by any assignment. But then, with all their gang-busting super-specialists, now and then federal law enforcement agencies would seek out my break-in services for special, sensitive, and critical investigations. There was the occasion, for instance, when a multimillion-dollar international drug-smuggling conspiracy implicated foreign diplomats and involved a platoon of federal narcotics agents, the Canadian Mounties, the U.S. Customs

Bureau, President Lyndon Johnson, Attorney General Robert Kennedy—and me.

In February 1964, I was assigned through the Police Commissioner to work undercover with the Federal Bureau of Narcotics (FBN). My contact, Jim Hunt, a New York FBN group leader, told me to bring my own burglary tools to a meeting where he'd brief me. The operation was extremely delicate; discretion was imperative.

At the time, I knew, New York suffered a devastating shortage of heroin, and a shortage there meant a nationwide drought. I figured this project must be related to an incoming shmeck shipment to relieve the junkie tremors. This was the picture I got from Hunt and George Belk, the Special Agent in charge of FBN's New York office:

About three and a half years ago, FBN nabbed Mauricio Rosal, the Guatemalan ambassador to Belgium and the Netherlands, for smuggling in 224 pounds of narcotics.

"We have another one," Hunt said, "another bastard with diplomatic privileges walking freely through airport terminals all over the world without customs inspection. We think this new fellow will be coming in dirty tonight from France—under diplomatic immunity.

"Some time after collaring Rosal I picked up a piece of information from a European source that there's another ambassador active in dope-running. A real heavy hitter. Lots of kilos and cashing in a thousand bucks for each kilo brought into this country. The stoolie told me this 'dip' had a code name, LAMBO, and he'd been in the racket for at least ten years, enlisting Rosal and others. When we checked with the Sûreté in France—that's where Rosal came from—we were told, yes, they had heard about a smuggler-courier named LAMBO but could give us nothing more.

"Since then a hell of a lot of work has gone into this and now we're sure. We know that in Nice a suave, arrogant guy we've been watching for months—a Salvador Pardo-Bolland—recently met with a big-time junk dealer, a Frenchman named Bruchon. LAMBO is an abbreviation of L'Ambassadeur Pardo-Bolland. He is our man."

As the case developed, this was to be the cast of bad characters:

Salvador Pardo-Bolland, fifty-five, career diplomat, Mexican ambassador to Bolivia. Suspected of having had a busy hand in narcotics for the past ten or fifteen years. Associated with a big organization, he recruited:

Juan Aritzi, sixty, employed by the Uruguayan Foreign Ministry in Montevideo. A master courier ("mule") with a diplomatic passport.

René Bruchon, fifty, a French citizen. Pardo-Bolland's connection with the American mobs. Listed on FBN's rolls as a "major international trafficker" and a representative of:

Gilbert Coscia and Jean-Baptiste Giacovetti, kingpins of an international ring, operators of the largest heroin manufacturing plant in the world. Linked to Mafia families in the United States.

Recently, Pardo-Bolland, under close surveillance in Europe for some weeks by FBN agents, had entered France from Holland with a false diplomatic passport. P-B, as we were to call him, had rented a car and spent much time driving through the French countryside, often visiting churches. Our men tailing him speculated that he might be receiving instructions by way of messages left in particular pews.

The Mexican drove to Nice where he was observed huddled with Gilbert Coscia and Juan Aritzi. They gave him two brown suitcases. Then P-B flew to Paris. Aritzi, according to his shadows, went on to Marseilles, acquiring two more suitcases.

The Sûreté and FBN agents assigned to follow Pardo-Bolland described him as "very wary . . . constantly looking for a tail." As a result the surveillance loosened up in Paris, enabling the itinerant Mexican envoy to give them the slip.

Through our sources we learned that a huge shipment of narcotics was long overdue in the States. Lacking enough merchandise to keep their dangerous customers calm, pushers in New York were already fearful of being seen in the streets. Clearly the heroin now had to stream in uninterrupted.

Pardo-Bolland had long been riding high, and over the years the Corsicans had paid him off generously. Now he was on the spot. So though he sensed he was viewed as a hot suspect, the Mexican was bound to be under heavy pressure to go through with the run. In New York the FBN was staking out an ambush for him on all flights from Europe.

"That's about it," Jim Hunt told me that day. "Your job, Mac, is to get into the ambassador's bags, find out what's in them. We can't just brace him, step into him. If we do and he's clean, all our work goes down the drain, he gets away with years of bringing in dope. And we have one big diplomatic incident."

"Swell," I said. "P-B isn't the only one under pressure."

Hunt explained that although the ambassador's movements that day were unknown to us, he did have a reservation for a flight from Orly Airport to Kennedy due just after 9:00 that evening.

"We don't think he's on this flight but when he does come, this is our plan: While his plane is taxiing up to the terminal you'll be moving along with it in a small panel-body truck. Customs will drive it right under the plane's wing. As soon as it stops rolling, the Customs guy and our agents will open the luggage section in the belly and toss the ambassador's bags into the back of the truck to you. Now, Mac, this is sticky as hell; you'll have only seven, maybe twelve, minutes to open, search, and relock four or five suitcases."

"Hey," I protested. "I can't guarantee. These will probably be foreign-made bags and each of them may have three or four locks. If the suitcases are clean and I don't get them all relocked in time this thing is going to blow up."

"What else can we do? This guy's got diplomatic privileges, V.I.P. treatment and no customs. The Mexican will expect to be in a cab ten minutes or so after the plane lands."

His Excellency, the Mexican Ambassador to Bolivia, did not show up on the nine o'clock flight from Paris.

We set up a temporary base in an unused Customs office. Bothering all of us was the limited time I'd have for my lock picking.

Friday night and all the next day Hunt and I kept vigil. Sunday morning our agents in France informed us that P-B had been located and was ticketed on a TWA flight due at 6:00 P.M. Our Narc squad congregated. Every one of us seemed uptight at the prospect of hitting an ambassador. I was left strictly alone, like the star of a show.

At 5:30 I stepped into the rear of the Customs truck, along with a couple of agents. The TWA jet landed. Following our scenario, as it taxied to the terminal our truck stopped under one of its wings. The agents tore open the baggage compartment.

At Orly, orders had been issued to make sure the Mexican's luggage was to be loaded last, ostensibly to facilitate its handling in New York, a bow to the ambassador's status. But it was fifteen or twenty minutes before I could tackle the diplomat's European bags in the truck. As I'd guessed, they were ornery—four locks, two on top, one on each side. Rushing, I grabbed the first bag, standing it upright to pick the locks. Opening it, too late I realized the wary smuggler had set a trap for any inquisitive narc. He had stacked paperback books in such a way that if the suitcase were not opened properly they would fall, breaking the pattern in which they'd been packed. He had also placed paper matches around the rim of the suitcase—and they dropped to the floor. Otherwise the bag was stark empty!

Reconstructing the traps as best as I could, I relocked the suitcase, then opened and relocked the other four pieces of luggage with greater care. All I found there were normal diplomatic contraband—some bottles of perfume. And clothes.

Not a sign of any dope. Obviously this was a dry run. The devious diplomat was testing the wind.

After P-B checked in at the Hotel Elysée on the West Side he remained in his room for an hour and a half. The Feds contrived to take over an adjacent suite and establish a radio base station. When he left I picked the lock on the Mexican's room, stepped in, and slipped the end of a paper clip into the lock to prevent use of a key. The paperbacks were piled on the floor next to his bed. In an ashtray I found the matches from his luggage trap.

The agents tailing Pardo-Bolland on the streets used their walkie-talkies to report to our hotel base station that he was strolling casually away from the Elysée. In his room, where I was joined by two agents, we examined the contents of his bags more closely. All we found was a white cushioning material lining each of the suitcases—similar if not identical to the kind used years ago by Mauricio Rosal on his smuggling ventures.

An agent came in to tell me that some FBN technicians wanted some samples of the packaging material and other stuff. I let them take it.

A message reached us that Pardo-Bolland was meandering

around Grand Central Station, looking at lockers. When the technicians started work I warned them not to disturb the stacked books and the matches. After a half hour, the lab guys gone, I made sure everything looked as it had when I broke in. Emerging to the corridor I slapped the face of the lock cylinder with a small magnet, thus removing the piece of paper clip from the keyway. I relocked the door, turned as the elevator doors opened on the floor—and for the first time faced the Mexican ambassador.

Trying to appear nonchalant, I walked toward him, nodded as I passed, and pressed the elevator's Down button, not daring to look back. Did he? I could hear him opening his door. In the elevator I turned toward the corridor. Our eyes locked for an instant before the elevator doors shut off the encounter. I sank into a sigh.

Later that evening Hunt filled me in on what Feds tailing Pardo-Bolland had learned. Leaving Grand Central he had visited a cable office where he had sent a message to Holland: "Arrived in good health." Evidently the diplomat was satisfied his dry run had been uneventful; after checking his paperbacks and matches, he was cool. Still, the live heroin shipment had to come in. Was the cable a signal to start the real game, that the way was clear for the actual delivery? Who was to bring in the goods? What happened to Juan Aritzi, the Uruguayan courier?

That night, while P-B slept blissfully, an alert woke us at the Elysée: Aritzi had embarked on a flight from Paris with six pieces of luggage—to Montreal.

The run was on.

Next morning the streets around the Elysée were flooded with narcs, each trying not to look the part he was playing and doing well at it. In our suite the base-station radio was busy with news from their walkie-talkies as people emerging from the hotel were photographed and followed until the Feds felt satisfied those citizens were not involved. Whenever Pardo-Bolland sauntered out, I'd be tipped off: "O.K., he's out, pop it." "Pop it," to the agents, meant what I did with locks. I'd go through the ambassador's wastebasket, check his suit pockets, look for anything to further our knowledge.

In Montreal, Feds contacted through Washington converged

on the airport to coordinate the operation with the Canadians, the RCMP.

When Juan Aritzi finally arrived, he put four of his six suitcases in terminal lockers, then registered at a hotel. The FBN and the Mounties, acquiring master keys for the lockers, shortly after midnight removed the luggage to the terminal office. Early on they had enlisted and alerted several reliable Montreal locksmiths who proceeded to open the suitcases with a variety of keys.

In the four bags the agents pounced on 61 kilos (134.2 pounds) of 80 percent pure heroin, packed in one-quarter kilo polyethylene bags and cushioned with white material. This dope was the best, the tan Turkish product, commanding a street value ranging from $13.5 million to as much as $50 million, depending on the demand.

Now a dilemma struck the lawmen. Making the seizure, the Mounties could wrap up an airtight case against Aritzi—thanks to the initial efforts of the U.S. Bureau of Narcotics. But at this juncture the United States could pin down little evidence against Pardo-Bolland, the prime mover and organizer of the smuggling caper. Urgent calls from Washington to Ottawa shaped up an agreement: The Canadians would allow the courier to continue his run into the United States, reserving the right to prosecute him later.

What could, or should, be done with all that dope? The heroin in all but four of the polyethylene bags was removed. Each of these four bags was marked and initialed by an FBN man and a Mountie. One heroin bag was placed in each of the four suitcases for future evidence. The Mounties kept the rest of the contraband, then accumulated 129.8 pounds of buckwheat flour from their commissaries and packed it into kilo bags (quarter-kilo bags were unavailable) for the suitcases. The relocked luggage was returned to the terminal lockers.

In the morning Aritzi, retrieving his suitcases at the terminal, appeared so nervous that he had trouble inserting his locker keys. He asked a porter for help; the obliging porter (a Mountie) served as another link in the chain of evidence being forged.

Flashing his diplomatic passport, the Uruguayan breezed

through customs. The train he took to New York probably carried more federal agents per square foot than were loitering around the Elysée. The strategy was to let the abbreviated heroin freight go through as far down the route as possible, even to somebody's kitchen table, while quietly pulling in Pardo-Bolland. Possibly, too, something of evidential value would develop that could tie the diplomats' suppliers—the Corsicans—directly to the load.

For this vital ultimate objective, as the stakeout heated up, Commissioner Charles Guillard of France's Sûreté had flown to New York. Since he was well-known to the drug trade, the Sûreté chief was put up at the Americana, away from the hub of the action at the Elysée.

On the day Aritzi embarked on the New York-bound train from Montreal, all his suitcases crammed into his compartment, the Americana Hotel oddly appeared in the picture. Pardo-Bolland in New York (his phone tapped, of course) called that hotel, asking the desk for "Mr. Rouge." No, Mr. Rouge was not registered. By now every narc in town knew that René Bruchon was probably the connection between the diplomats working for the Marseilles manufacturers and the U.S. mob. We figured "Rouge" would turn out to be Bruchon, and every agent carried his photograph. Jim Hunt opened a branch office at the Americana, occupying a suite on the forty-ninth floor.

Aboard the train, Aritzi never took his eyes off those suitcases. He arrived at Pennsylvania Station at 9:00 in the morning.

I had been awakened at 6:00 A.M. for a briefing, along with some thirty agents, and we all descended on Penn Station. There, Aritzi's suitcases were unloaded by a black Fed acting as a porter. Another agent, a "Navy photographer," shot movies of the diplomat while pretending to take pictures of his flower-laden "wife." The films showed Aritzi diligently supervising the placing of four of his suitcases on a hand truck, then accompanying them to the baggage checkroom for storage. His remaining two suitcases were taken to a taxi—driven, naturally, by one of Hunt's men.

Armed with a court order, Hunt, Special Agent Belk, and I invaded the baggage room. Picking open the four suitcases, I quickly found the initialed bags of true heroin.

"I swear this stuff is just like Rosal's," Hunt said. "And the cushioning here is like what we saw around the perfume in Pardo-Bolland's luggage."

"We'll take a sampling," said Belk, "but our job now is to get Pardo-Bolland—and Bruchon—so tight that we won't need this kind of thing. Getting them right will give us our best shot at the Corsicans, too."

Some of the heroin was removed as evidence, we added our initials to the four incriminating polyethylene bags, and I re-locked the suitcases. Back in our car, the radio told us that Aritzi had checked into the Elysée. Incredibly, his room adjoined ours. So we were sandwiched between P-B's room and Aritzi's.

Five minutes after Aritzi checked in he barged into the Mexican ambassador's room. Through our walls we could hear them jubilantly greeting and congratulating each other. Again P-B phoned the Americana, inquiring about "Mr. Rouge," and again received a negative response. Their voices dropped to whispers. Now the team was troubled: The heroin was in, they had the baggage claim checks, but no one had appeared to receive the goods.

We, too, were worried. We needed more to guarantee ensnaring the ambassador, and stateside we had nothing on the clever Bruchon. None of us could understand what might have gone wrong. Belk phoned Narcotics Commissioner Henry L. Giordano and Attorney General Kennedy in Washington.

We were looking for soft spots. Had there been a screw-up in Europe, where the Corsicans were being watched? Maybe a leak? A tip-off to the shippers? Had the Corsicans—or the American dealers—become aware of something fishy? Had I left behind any trace of my visits to P-B's room? From the way the diplomats had rejoiced on a faultless delivery so far, they felt cool, so that wasn't a giveaway.

Aritzi and P-B, out to dinner, cabled Carmen Lopez (a Coscia alias) at a hotel in Cap d'Antibes, France: "Cousin cannot be located at indicated address."

All the next day the diplomats wandered at random, once to Penn Station where, from a distance, they stared blankly at the baggage checkroom. Their faces showed strain.

The long-awaited break turned up early the following morning when Pardo-Bolland was aroused by a curt phone call from France: "The name is Blanc. He is waiting." We guessed the Corsicans had taken a precaution, though P-B had faithfully accommodated them as a courier for years. Coscia had given him an initial false contact, just in case the jittery Mexican fouled up and was right about being "hot" as a suspect.

Exuberant, at 8:45 the ambassador and Aritzi hurried to the Americana. While they were away I hit P-B's room. On a sheet of hotel stationery beside the telephone table I found the word "BLANC." The paper was photographed, and I left it behind.

At the Americana, Pardo-Bolland and Aritzi asked for "Mr. Blanc" and went up to Room 4924. They were greeted by a fidgety and anxious René Bruchon, their pickup man.

Hunt, at the Americana, reached me at the Elysée, informing me that Bruchon was right across the hall. "He's been there for days! Come on over. We'll need you to pop the connection's room as soon as Bruchon goes out."

After a short stay with Bruchon, the two diplomats left and promenaded along Fifth Avenue, plainly delighted with life. They studied gleaming American cars in showrooms and then dined leisurely.

When Bruchon left his room, I had my first crack at it. In a single-lock suitcase I came upon a schoolboy-type notebook. At least a dozen pages were covered with figures and letters indicating no clear pattern. Obviously a code, maybe a record of shipments and contacts. Before I relocked the suitcase I made note of the four suitcase keys I'd found in it. They were on a tasseled string and none fit Bruchon's own luggage. No baggage checks.

Clearly the next move was Bruchon's; he'd have to get in touch with the New York buyers. Moving a massive amount of heroin had to be accomplished with utmost care, not only to evade law enforcers but also to keep it from being hijacked. And connections like Bruchon have been kidnapped before by Turks (outlaws) within an organization. When a Bruchon type makes his play it's done quickly, the contact with the buyers held to a minimum. Ready for the critical juncture, undercover narcs rarely involved in overt investigations were brought in to cover Bruchon thoroughly.

By 9:30 that Thursday morning Bruchon was busy carrying out his mission, patrolling the blocks in the vicinity of the Americana. On Fifty-sixth Street the Frenchman became more intent, sharply eyeing passersby and, especially, cars driving by. Just before 10:00 A.M., tailing agents saw a white Cadillac convertible pull into the street, the second time within ten minutes. In the car sat two tough-looking young men, swarthy, probably Italian. Stopping at the curb they motioned to Bruchon to approach them. He leaned into the open window at the driver's side for perhaps a minute, chatted, then turned and walked briskly back to his hotel.

Agents rambling with their "lady friends" and others hovering in cars, pickup trucks, and government-owned taxis were straining at the leash. The stage seemed set: broad daylight, on a busy thoroughfare. Enter: the New York mob, in a brief encounter. When the Frenchman leaned into that car something must have passed, if only information, but nothing that could have been a sample of the shmeck; the drugs were under guard at Penn Station. No matter what it was, we sensed that we were now coming close to the kitchen table where the narcotics would be pawed over by the kilo distributors.

As the Cadillac pulled away it was followed by four of our vehicles, two of them taxis. Over their walkie-talkies the Feds' communications were strident, a jumbled clamor about their positions and how the subjects were moving. Jim Hunt could be heard booming above the din, directing the surveillance, selecting the men to take the "point" (the first car behind the quarry).

The white convertible drove east, turned onto the East River Drive, and headed downtown. Watching their rearview mirror for a tail, the strangers slowed down to fifteen miles per hour and let traffic pass them. The man in the passenger seat kept looking out his window for any suspicious car behind them. It meant our point had to be continually changed. Several times Jim Hunt drove ahead of the Cadillac, tailing by his own rearview mirror, exiting, and then returning to the Drive behind the target.

Cutting out of the Drive, the Cadillac rolled slowly through East Side streets, making one turn after another. Paused at a traffic light, the men let the light go green, red, and green a second time, then moved on. No doubt they had "made" our tail.

Suddenly accelerating, after a while the car stopped abruptly at a corner, pulling to the curb as if to park. The driver was seen dropping something into a sanitation basket and the car sped away.

Hunt, via radio, ordered one of his tailing agents to stop at the basket and search it. The narc dug out four suitcase keys on a tasseled string. Racing west, the Cadillac passed through red traffic lights, and briefly the trail was broken—time enough for the hoods to abandon their car and vanish.

The four suitcase keys, rushed to Penn Station, fit the luggage Aritzi had stored there. I had last seen those keys in Bruchon's suitcase at his hotel room. We believed they had been handed to the hoods as a way of verifying Bruchon's identity to the New York syndicate awaiting the shipment; the buyers probably had duplicates to compare with Bruchon's. The missing baggage claim checks—the most damaging evidence—might still be in the Frenchman's possession. Hunt beefed up the coverage at Penn Station. We couldn't be sure the claim checks hadn't been passed along with the keys.

About an hour after the two strangers disappeared, agents at our Elysée base station could hear Pardo-Bolland's telephone ringing. Both he and Aritzi were still taking in the Big Apple sights and celebrating, at lunch, another big score.

Bruchon, too, had slipped out of his hotel room, his insistent telephone unanswered.

The game appeared to have run its course. The next move was ours. The time had arrived for the bust. It was particularly important that Bruchon be yanked off the streets; otherwise his talents, experience, and contacts would permit him to slide into an underground system out of our immediate grasp. (We knew by this time that Bruchon had entered the country on a tramp steamer arriving in Baltimore.) Washington was briefed that we were about to make the bust.

A complication, however, a serious impasse, held us up. Just at this time the president of Mexico, Adolfo Lopez Mateos, was visiting President Lyndon Johnson and former President Eisenhower at the latter's Palm Springs home in California. Certainly the shocking arrest of a Mexican ambassador for smuggling narcotics—while the presidents were socializing and practicing hands-across-the-border amity—would prove mighty embarrassing.

We were undecided as to whether I should again break into Bruchon's room for that mysterious notebook.

"If he gets the phone-call alarm," said one agent, "he'll clear out everything."

But Hunt felt we had to wait until we got the green light from Washington.

The awkward quandary brought to our scene George Gaffney, second in command to the Federal Commissioner of Narcotics. He was on the phone constantly from the Americana to Washington. Like Gaffney, Attorney General Kennedy recognized the risk of delay in making the collar. Agents surrounding Pardo-Bolland and Aritzi moved in even closer. The Sûreté was advised to tighten surveillance over the Corsicans, Coscia, and Giacovetti.

At Palm Springs, LBJ was brought up to date on the case. Washington relayed to our Americana suite the President's request that we hold off the arrests as long as possible without jeopardizing our position for conviction. Gaffney was assured that LBJ would cut short his meeting with the Mexican president, while still observing the amenities. We assumed that LBJ would apprise Lopez Mateos of the situation.

Meanwhile the two dope-diplomats and their French connection, though not in each other's company, spent the rest of that day away from their hotels, thus frustrating their telephone callers. Bruchon was the first to return to his room. From our suite, sprawled on the floor, we could look under our door directly across the space under the Frenchman's door and see his feet. At 9:00 P.M. he went to bed. At 11:00 his phone woke him: a few crisp words warning him to flee.

Jim Hunt and I, our cheeks pressed to the carpet, peered under our door. Bruchon, in bare feet, kept pacing up and down the length of his room, tigerlike. His frantic performance reflected his terror—a trapped animal in panic. We could hear the toilet being flushed several times. Nerves? Or was he destroying evidence? It also occurred to us that the Frenchman might dive out of his forty-ninth-story window. Checking our own windows we were relieved to find they were of thermal glass, sealed.

Finally, at about one in the morning Bruchon dressed and, with his lights still on and a Do Not Disturb sign on his doorknob, left the hotel. I picked his lock and scoured the place like a burglar. No claim checks. The notebook I had last encountered in his suitcase lay on the night table, the coded pages torn out. He had taken only his hat and coat.

Back in our suite I found Hunt, Belk, Gaffney, and others huddled around our base-station radio. Excited agents were reporting Bruchon was acting like a madman, half-trotting through the streets and blindly bumping into people. He took a subway train uptown then back, scrutinizing people near him, waiting—as we were—for the arrest. Then he trudged into his room.

Undoubtedly Bruchon's highly sensitive antennae spotted the narcs all around him. He must have realized he couldn't go near anyone who might slip him out of the country. That would entangle others and he knew he was completely on his own now.

Pardo-Bolland and Aritzi, having returned to the Elysée at about 10:30, serenely retired for the night. Before dawn the Mexican ambassador caught his inevitable alarming phone call. Listening for a few seconds, he started to say something but the phone had gone dead. (We soon learned the call, originating in Switzerland, had ordered him to "go home.") The no longer suave ambassador, whisking from his room in his pajamas, burst in on Aritzi, leaving the door partly ajar. He shouted in Spanish, repeating the word "*casa . . . casa.*" The door slammed shut, their voices hushed. They stayed together for the remainder of the night.

Still no go-ahead from Washington.

In the morning the tension of the night before showed clearly in the haggard faces of the two conspirators as they emerged from the Elysée. I picked the locks on their doors for the last time. Walking into P-B's room I tripped over a clothes hanger just inside the doorway. Next to it lay the scarf he had always worn. His hat and coat were gone—and the piece of paper with the word "BLANC" scrawled on it. Aritzi's room, too, was in disarray, highlighted by an open bottle of whiskey. No sign of his hat and coat. Relocking the hotel doors I cut out to join the Feds on the smugglers' surveillance.

On Fifth Avenue the notorious LAMBO dashed desperately from one airline office to another, trying to book a flight—any flight—out of the country that day. The gaunt sixty-year-old Aritzi followed, panting. After a couple of disappointments the pair shuffled out of an airline office apparently satisfied. While Pardo-Bolland seemed much relaxed, Aritzi—still wrought up—gesticulated excitedly as they walked to lunch at the Waldorf. By the end of the meal they were actually laughing.

Over walkie-talkies tailing agents told us at the Americana that the diplomats had booked a flight to Florida and then to Uruguay late that night. We assumed they had been holding their breaths all night, waiting for a fatal knock at the door. Now Pardo-Bolland was probably rationalizing: The urgent phone warning in the middle of his sleep must have been a false alarm; if anything had been seriously botched, wouldn't they be hauled in by now?

We waited. And waited, advised that the Mexican president's departure was imminent.

At about 6:00 P.M. Bruchon, closeted in his room all day, drifted out into the streets. At 7:30 he was standing glassy-eyed at the corner of Eighth Avenue and Fifty-seventh Street when Washington phoned Gaffney at our Americana base. Hanging up, he announced:

"President Lopez Mateos is airborne."

The coast was clear. The message flashed out to the walkie-talkies. Two tailing narcs tapped Bruchon on the shoulder. He wheeled, sagged, and wet his pants. One of the men taking him in transmitted in a loud incredulous voice:

"The Frenchman's got the baggage checks on him!"

We scooted to the Elysée for the two diplomats, on their way back to their hotel. Word shot out: "Don't take them on the street. Let them get into their rooms."

I stood just inside our suite, the door open a crack. The two dope-runners ambled down the corridor past us and entered Aritzi's room. A moment later, from our open doorway, we watched Jim Hunt do the honors. Accompanied by a federal prosecutor and an FBN group leader, he placed LAMBO and "the mule" under arrest.

As expected, P-B and Aritzi, displaying their credentials, claimed diplomatic immunity. They were held but not arraigned until after midnight, to give the Sûreté and FBN agents in Europe a chance to grab the slippery Corsicans, Coscia and Giacovetti. According to a peculiar French law, a warrant for any committed crime could not be executed after dark. Though the Corsicans had been under close surveillance, a dawn raid in Marseilles failed to net them and they were being sought throughout Europe.

Next day, Mexico announced that Pardo-Bolland was fired because he had absented himself from his post without authorization, and diplomatic immunity did not apply to him. Uruguay disowned Aritzi on the ground that he had gone to Canada as a private citizen "for his health," that he had resigned his post before the trip, and therefore could not claim immunity.

Sensational headlines hit the press. The *Daily News* frontpaged it: $13.5 MILLION DOPE SEIZED;U.S. RAIDS BAG 2 "DIPLOMATS."

At the arraignment in federal court Bruchon looked uncomfortable in his stained trousers, a trembling Aritzi remained mute, a smiling P-B was the only member of the team who would speak English. When the charge of smuggling *two* kilos of illicit narcotics was read by the clerk it flabbergasted them. Not until later did they learn of our buckwheat ploy. The court was informed of their chronic involvement in drug traffic and of Bruchon's record: a conviction in 1949 for smuggling twelve kilos of heroin into the United States, conviction in Algiers on narcotics charges in 1956, illegal entry in Canada in 1959, illegal entry now in the United States.

(Pardo-Bolland, the narcs estimated, had been earning about $80,000 for each mission he undertook. Carrying on since 1954, making at least two trips a year, he must have scored a total of $1.6 million.)

Each of the three culprits eventually was sentenced to fifteen years in the federal penitentiary.

Announcing the arrests, Attorney General Kennedy commented diplomatically:

"U.S. authorities have worked with the closest cooperation of the government of Mexico, which has played a strong part in the international effort to suppress the illegal trade in narcotics." He added that the arrests had led to the disruption of the Coscia-Giacovetti manufacturing operation, then the world's largest producers of illicit heroin. Smashing the ring meant that at least three-fourths of the dope flow into this country had been damned.

Kennedy, Commissioner Giordano, all of us who worked so hard on the case did not fully realize the stunning impact the bust would provoke. With the choking off of most of the dope imports, an acute famine struck New York, Chicago, other cities. A bloody war broke out among the Mafia families struggling to lay their hands on what little heroin was obtainable. Street prices soared. Hard-up hopheads openly attacked people for money. Pushers, to get their rampant customers off their backs, handed them "hot shots" (rat poison). Junkies were found dead in their furnished rooms or plush penthouses. There was, as one writer later put it, "panic in needle park."

One galling question hung over us: Who were those two young sharp Italians? We never found out.

I still find it hard to understand why so few people pay little attention to the ease with which even a semiskilled thief can go through locks with picks. Even a Pardo-Bolland, despite all his precautions, let me take so much from him so easily. Good combination locks on his luggage would have slowed me down considerably. A portable alarm in his room could have given me fits. If he and Bruchon had kept their vigil in hotels using the most highly pick-resistant cylinders available—and if Aritzi had observed similar vigilance and carried his valuable baggage with him—we could well have been completely stymied.

My favorite aphorism: *The pick man, like the devil, has as his strongest weapon the fact that people just don't believe he exists.*

Chapter 10 After a burglary

In Neil Simon's play *The Prisoner of Second Avenue,* the heroine comes home one afternoon to find her apartment all but annihilated, thoroughly plundered. She flies into wild hysterics. Few in the audience titter; most people share her shock either because they, too, have suffered a similar experience or because they dread it.

A real-life victim, Mrs. Wilma Tremaine of Chicago, was bitter after she entered her purloined apartment. "The loss of our good stuff," she said, "didn't bother me as much as the thought of a stranger going through my home. I felt so vulnerable, so violated. It took me weeks to calm down."

Recently, Albin Krebs, a *New York Times* reporter, was stunned by the seventh burglary of his apartment in twenty years. In the latest break-in, thieves had even torn family pictures from frames, looking for money, and one had left a half-eaten apple on the dining room table.

On his fifth burglary—the "worst"—Mr. Krebs had arrived home from his father's funeral and encountered two burglars on the stairway.

"I was emotionally drained by grief," he reported in the *Times,*

> and I didn't use my head, for foolishly I blocked the stairs with my suitcase. The burglars attacked, tore my sports jacket off me, and ran back up the stairs to the skylight. I followed, and one of them threw a garbage can at me and struck me on the head.
>
> This was the burglary that caused my insurance company to drop me. The companies that write those policies refuse to renew them after you've been burglarized a couple of times and put in claims.
>
> How does it feel to be burglarized? You are left with a permanent sense of dread, an irreversible feeling that your home is no longer your own private place. It has been invaded, raped, so to speak. Your

> most personal papers and photographs have been handled by vicious hands.
>
> But most of all, you never feel safe again. The latest burglary brought back all of the daily, nagging fears. Each day, as I climb the stairs of my apartment, my steps are measured, my eyes wary. Outside my door, as I am about to insert a key, I imagine I hear a sound inside. Can it be another burglar? Did he come in through the window? Did he hear me on the steps?
>
> Sounds melodramatic, doesn't it? But ask any other burglary victim about that, and he'll tell you his own story of the fearful and brutal souvenirs from burglary, a crime against people.

It's not at all uncommon for distraught burgled citizens to leave the scene and check in at a hotel or motel. Their home doesn't feel the same any longer. Then they're prone to overreact, marching off to some discount store to buy and install a lot of useless hardware on their doors that will not be effective except to trap them inside in the event of fire.

Some people are inclined to take the next day off from work to shop for a watchdog. They may call the police to inquire about a gun permit or sleep with a baseball bat under the bed. Gradually the shock wears off and the only time they think about the incident after that is when the subject comes up over cocktails. Then they probably remark:

"We were really upset over it at first, but what the hell, it's the world we live in. You can't keep these bums out if they really want to rip you off. We had lights burning and everything."

So in many cases, the residence remains just as vulnerable as it was before, the brief flurry of activity in buying cheap chains and die-cast locks at a hardware store serving no real purpose. A month—or a year—later if he is hit again our householder is more convinced than ever that "if a burglar wants to get in he will get in."

The emotional impact on people in an invaded home is well appreciated by attorneys defending criminals in burglary cases. Occasionally a prospective juror is asked, "Has your home or the home of any friend or relative ever been burglarized?" An affirmative answer will almost certainly result in the juror being excused "for cause." After repeated rejections of prospective jurors on these grounds, an exasperated judge exclaimed:

"Counselor, if we are going to continue to include 'friends' in that question, I fear we will never be done with this jury selection. I would suggest we limit the field to the juror himself and his immediate family."

Pity the "poor defendant" appearing before a judge whose home has been burglarized. A jurist who presided in Brooklyn courts for many years had his apartment defiled five times. In three of these break-ins the thieves defecated on his bed. More burglars were convicted in his court than in any other in the state. But, alas, the judge also had a dismal record for reversals of these convictions, based on his trial rulings, which consistently went against the defense.

Few victims are immune to these typical reactions. After his home was plundered, a former governor once confided to McDermott over a cup of coffee: "You know, Detective, I don't want you to leave. When I was in office I had bodyguards, but I never felt their need as much as I do tonight."

It turned out that this burglary was committed by a young thief brought home by the ex-governor's niece after she met him in a midtown bar earlier in the evening. When she went to the bathroom her guest dashed into her uncle's bedroom and fled with his pockets stuffed. McDermott, finding no break and the locks not picked, got the young woman to admit to her embarrassing pickup. The thief was later identified, but hard evidence was scanty. The ex-governor was advised by an assistant district attorney not to prosecute, "for the sake of the young lady."

When you discover a burglary

On returning home, if you find exterior signs of a break-in—an open door, a broken window, an obviously forced lock or one hard to get a key into—don't enter, and refrain from calling out. Go to a neighbor and phone the police. If you have entered before discovering that you've been ripped off, don't touch anything. Touching can blur fingerprints the burglar may have left and these could help police tag the criminal. (But don't count on cops bothering with investigating for prints in a run-of-the-mill burglary.)

Even in burglaries considered of importance, experienced detectives have little hope that dusting for prints will accomplish anything more than messing up your ransacked home further. It takes a rather deliberate placing of a finger for a thief to leave a really good print. Identifying a thief from just one print, no matter how good it may be, is a very remote possibility—unless police pin down a suspect. Most prints lifted by technicians at a crime scene are "partials," offering only a few identifying characteristics. The law requires a very specific number of "identifying points," and these are not often evident. Few surfaces in your home are conducive to leaving a readable fingerprint.

Note that insurance companies are increasingly loath to issue burglary policies. As a member of one major insurance brokerage firm put it:

"The public must take a portion of the blame for crime losses. People think, what the hell, my two televisions are insured; so they don't bother to lock up or take any of the precautions they should when they leave the house. The industry can't be expected to make up for people's losses that are the result of carelessness, but if the public isn't interested in security it shouldn't expect the companies to subsidize such lack of concern."

Even if you believe, cynically, that police are unlikely to solve the crime, be sure to notify them. It's estimated that nearly half of all residential burglaries are never reported. If you expect to collect on insurance, there must be a police record.

Indicate to the police any possible suspects—such as a new building employee or a recently fired domestic. Women tend to be particularly good at supplying police with leads to suspects. Not only are they more familiar with the home area but they do have a keen perception of the oddball males they may encounter.

Time after time a housewife somewhat reluctantly has told an investigating detective about "a delivery man who makes me feel ill at ease" or "just the other day the new building superintendent's son was wandering through the hallways—don't laugh, but I don't like his eyes." Women in a building or neighborhood, when asked about their feelings after a burglary, will describe this or that local resident, canvasser, or repairman, and a good detective pays attention when the words "surly," "creepy," "sly," or "sneaky" are used. With only such remarks as a starting point, a lot of loot has been recovered by police.

Always give the cops a good description of the property stolen. Compile a list of what's missing; look everywhere, in closets, attic, basement. In almost every case victims will telephone police long after the crime to report additional items stolen, those not immediately missed. In most instances this doesn't disturb the police, but now and then the items not initially reported are those most readily identifiable. A day or two after your loss the police may be investigating a suspect for an entirely different matter; on his person, in his car, or at his home they may have seen, wondered, and questioned him about articles they surmised were loot, but could do nothing without a full report of your loss. The credit card not reported by the young couple in the Braddock murder case (the "climber") stands out as an example of an even more important reason for complete reporting.

So be sure to give the police a copy of the list you compiled, including any serial numbers you may have for missing articles. Besides proving your ownership, this could help in retrieving the stolen property.

The police should have the names of all persons who have duplicate keys to your residence. Inform the investigators of any strange phone calls or suspicious visits by individuals representing themselves as door-to-door salesmen, delivery men, or pollsters. Posing as one of these, the thief may have cased your home before returning to burglarize it.

Detective McDermott recalls being often discouraged by police files on stolen property.

"A burglar is arrested. I'm armed with serial numbers I copied during a 'visit' to his pad—numbers taken from half a dozen TV sets, cameras and hi-fis I found there. Then I'm confronted with a file listing hundreds of Minolta 101's, RCA television sets, and other loot with no serial numbers, as well as sound equipment where the victims could not even supply a brand name. I have the stolen stuff but I just can't give it back to anyone. What happens to it? Well, police departments hold a periodic auction of property seized or found by them that can't be identified.

"Still, every once in a while there'd be that lovely card in a police file listing a serial number. Maybe with a photo of a jewel in its setting, complete with a jeweler's 'charting' of the diamond —a diagram of carbon spots (clouds), fissures, and 'feathers'— as good as fingerprints to me."

Your insurance

As soon as possible after notifying police, call your insurance agent to inform him of the burglary. Remember: If a newspaper report has the police investigators term the crime "a mysterious loss"—a phrase used by many insurance men—it usually means the victim, if insured, will never collect on his policy.

If you have not used the best lock cylinder available (to rule out picking), can find no evidence to establish how the crime was accomplished, and you're sure that your doors and windows were locked, ask the police to remove your lock cylinders to determine if the burglary was the work of a pick man. You have to persist because few police departments have officers qualified to make a determination. They should, however, be able to send the cylinder to the State Police Laboratories.

Search everywhere for marks of entry. You may, for example, spot scratches on a doorjamb from a jimmy or notice a broken window. Show such signs to the insurance adjuster when he visits you. Most companies will not pay if there are no indications of forced entry; the adjuster will conclude that you probably left your doors unlocked—and that constitutes contributory negligence.

Of course, give the adjuster a list of your stolen property on the forms he leaves with you for filing your claims. Before filing, collect all the information the company requires, including the price you paid at purchase and the evaluation of each item or its replacement value.

Insurance companies naturally make no allowance for sentimental value. When comedian Jerry Lewis was ripped off for thousands, his main concern was for a ring valued at only forty dollars. This was also true of Mrs. Beatrice E. Fleming after her million-dollar burglary. When police gleefully told her that all her property had been recovered and the culprits collared, she ignored everything in her quick review of the property, except a ring worth nineteen dollars that her son had given her years before.

As much as possible, you should have tangible proof of ownership or value of your possessions. In one case a couple had nothing to indicate the value of a portable chest of silverware they had received as a wedding present long ago. They knew neither the manufacturer nor the design. The insurance company settled for the value of a cheap replacement set minus one-fourth for its use over the years (depreciation). So it's smart to keep a receipt or a canceled check on valuables you have bought to establish a claim under the theft section of a homeowner's or tenant's policy. On gifts, of course, this is not practical.

Many burglary victims are made unhappy when settlements are based on policy ceilings they were unaware of. Most standard policies have a $500 limit on which the company will pay for furs, jewelry, and stamp collections. On cash the limit is usually $100. You can, of course, raise the ceiling by buying a floater, at additional cost, that is tailored to specific articles like expensive cameras and art.

As for deductibles, on many conventional homeowner policies you have to absorb the first $50 of loss yourself. If your home or apartment is particularly vulnerable you can opt for a higher deductible. Still, insurance seldom compensates you entirely for losses. On certain items "depreciation" can run as high as 100 percent. Your stamp collection may be worth thousands, yet all you'll probably collect is $500, the limit on such an item. On silverware, guns, and gold objects reimbursement is now limited to $1,000 in most states.

Note: On stolen property not covered by insurance, your listed items should establish the loss for tax purposes.

The life of an insurance investigator gives him much cause to question public morality; fraudulent claims are commonplace. A police detective doesn't ask probing questions about the value of this or that article; he is far more interested in a good description. He knows that usually the value of stolen property is exaggerated by the complainant for tax purposes or as a cushion against what he feels the insurance company will settle his claim for. A merchant who has his safe cracked will automatically add to his dollar loss the cost of replacing his demolished ancient safe.

Some people believe there's great prestige attached to a spectacular loss in jewelry. Initial loss claims of actresses have been known to shrink from a million in gems to a mere $60,000.

One unfortunate musical comedy star lost her treasures just after renewal of her insurance was refused, although she had been paying premiums for thirty-five years. The lady had left her California home to appear in a Broadway show that looked like it would enjoy an extended run. She brought her most costly jewelry with her. When her policy came up for renewal the company turned her down; they would not insure as long as she was a resident of a New York hotel. While she was on stage one night, her suite was ripped off. If she had remained at home in California, she would probably still have been covered by insurance.

Some insurance companies are getting back into the burglary policy business. They are just beginning to realize that there are cylinders that won't be picked and locks that do "lock." These companies are not looking for reasons to decline renewal of policies; they are just insisting that you use what they require by way of protection. Then they are confident the burglar will avoid your residence—and go elsewhere.

Federal theft insurance

Yes, the federal government does offer residential and commercial crime insurance, at comparatively low rates, in nineteen states and the District of Columbia. It applies to places—high-crime areas—where coverage is hard to come by. Policies are sold through the same licensed agents and brokers who place insurance for private companies. But because of the low agent's commission relatively few brokers push the federal policy.

The policy has a $50 deductible or 5 percent of the gross value of stolen property, whichever is greater. You must be responsible for your own protection. All apartment or house exterior doors must be secured by deadlock or latch; all windows opening on a basement, first floor, or fire escape have to be fitted with a lock of some kind.

If you live in a high-crime neighborhood, if you're single or both you and your spouse are out of the residence, working during the day, federal theft insurance may be for you. Your insurance agent or broker isn't interested? For information, call this toll-free telephone number: (800) 638-9780. Or write to Federal Crime Insurance, P.O. Box 41033, Washington, D.C. 20014.

What happens to your stolen property

As we've noted before, a burglary is generally far more difficult to solve than a homicide or embezzlement, where witnesses or tangible evidence may point to a specific culprit. Be reconciled to the fact that the chances for return of your property are slim. Nevertheless, each year millions of dollars worth of stolen articles are recovered. If your missing items have been properly marked with serial numbers or your Operation Identification stamp, you may be notified—as in New York—by the Police Property Recovery Squad.

Under Operation Identification, as originated in California, your property is marked with your driver's license number, your address, or other identification. You display the decal the police gives you indicating that your property is "registered." It serves as a deterrent to prowling housebreakers.

Crude and amateurish thieves sell their stolen TV sets, tape recorders, and typewriters privately to individuals, often to secondhand dealers or pawnbrokers. These businessmen are required to maintain records of all property purchased, and their records can be a source of retrieval for police—but, again, only if the report on stolen goods provides a valid description.

Dealers who ask a seller for proof of identity will accept such credentials as a driver's license. In our society, thieves can obtain just about any form of I.D., counterfeited, for a few dollars. A criminal who knows the ropes can get a birth certificate, social security card, or just about any other document issued by government agencies involved—in any spurious name. But few of the criminals performing run-of-the-mill jobs go to these extremes. They simply use the utility bills, credit cards, or anything else stolen from a victim to identify themselves when selling your property. Wouldn't you think, though, that a pawnbroker seeing the same fellow come in day after day with similar goods might get suspicious? While the vast majority of people in these shops are honest, there are those who share responsibility for the crimes of their clientele.

Professional burglars use fences, or criminal receivers, who pay them "top dollar," which will be about one-third of the value for precious stones. A mink coat nets the thief about one-fifth of its value; stamp collections, 10 percent. Large or distinctive diamonds will generally be taken out of their settings ("put on the break") and the settings melted down. Your beautiful ring setting or bracelet of white gold means nothing in the burglar's world; he is paid only for the stones they house. Charm bracelets with their readily identifiable disks—often inscribed with personal messages—are much too hot for a fence to handle, so they are committed to the furnace. The thief receives pennyweight value for the gold, a pittance.

Bringing in good pearls, a break-in artist will be swindled by his outlet. The jeweler-fence knows that very few of his criminal-clientele know genuine from paste when it comes to pearls and so invariably the verdict is: "It's paste. Leave it with me and I'll try to find a sucker."

Manny, a veteran jewel burglar, insists he gets a much higher return than others of his ilk. Soon after a theft he passes his loot to several favorite fences, who remove the stones from their gold-and-platinum setting. In this way the gems, usually identifiable when mounted, can easily be mixed in and passed along with legitimate stones. Manny receives an "ounce count" on the gold and platinum settings for which he's paid additional cash. Some of the skillful fences are engaged in the legitimate jewelry business; others are free spirits roaming the world's markets and maintaining Swiss bank accounts.

Unlike most larcenists, Manny never unloads his entire score with a single stolen-goods receiver; it's his safeguard against the arrest or heist of the one fence. This also puts Manny into a better bargaining position. Cannily he never keeps any of his stolen goods as a souvenir or as a present to a lady friend.

Some years ago the FBI sought and won concurrent jurisdiction with municipal police in cases of burglary where the score was in excess of $10,000. The FBI took the position that loot of this magnitude usually meant that the property had to be transported across state lines, which placed it in their courtyard.

In New York City this attitude is unreal. New York constitutes the largest market for hot jewelry in the nation, if not the world. Burglars who work their trade in New York, and the fences they use, don't send the stuff out of the city. Smart crooks in other locales are far more likely to bring the jewelry into the New York market, where it just vanishes. Honest jewelers have told McDermott that unquestionably they have bought and sold stolen goods. Particularly diamonds.

As one man said: "Who knows where they've been—or how long they've been around. A lot of the big ones have been stolen at one time or another."

And one good jeweler observed: "The young man buying an engagement ring, he won't go to a secondhand dealer or an auction. He buys the girl's ring at a place with blue velvet and a fancy chandelier. There, he probably gets a stone that's changed hands four or five times. Ah, love is wonderful."

McDermott is reminded of some episodes involving slippery fences:

I know many of the jewelers in the New York exchanges and some I've locked up. Others I wanted very badly to put away but couldn't get near them. One still irks me: a fence I'll call Armand Bodeline.

It seemed that every time a good pick man was grabbed he would have Armand's card in his wallet. When we searched the crook's apartment I'd find Bodeline listed in his telephone index. The Bureau of Criminal Identification had nothing on Bodeline although detectives talked about him often when big scores went down.

Once, I heard that Bodeline had acted as an "anonymous intermediary" in the sale of a stolen diamond to an insurance company that had paid a claim on its loss. Now, this is a no-no. Insurance companies have no special privileges, so I went after them only to find that my information was wrong. The company had gotten the stone back, all right, but legally and not through a fence. The stone, a distinct outsized yellow, had been spotted in the showcase of a Forty-seventh–Street jeweler by a competitor who called the police. Identifying the stone, detectives seized the diamond and brought the dealer in for questioning.

He told police he had bought the unusual diamond three or four weeks before from a man he named, a Mr. K. Detectives picked up Mr. K. He insisted he had bought it from a woman in an East Side bar who said it had been her mother's and she was trying to sell it but had been offered only $3,500. They made a deal for $4,000, Mr. K. met her the following night, paid her and got the ring.

The next day Mr. K. went into the Forty-seventh Street jewelry exchange and sold it at his first try for $4,750. The woman could not be found. The jeweler received his $4,750 back from Mr. K. The police turned the stone over to the insurance company and one New Yorker—Mr. K.—learned not to buy jewelry from young women in bars. Incidentally, the insurance company eventually sold the stone for more than it paid out on the claim.

Armand Bodeline remained unscathed. We tailed him and had court orders to tap his phone. I "visited" his second floor office and went through his safe in the Bowery jewelry exchange. Nothing showed up. But I know he's the equivalent of a thieves' partner. Armand just outlasted me. Some day another detective will nab him.

What you should do

In the case we related involving Thelma and Mort Weisinger of Long Island, New York, among the valuables stolen were thirty-five recently purchased Consolidated Edison bonds worth about $22,000. The certificates had arrived a few days earlier and the couple had delayed lodging them in their bank safe-deposit box. Wisely, Mr. Weisinger reported the loss to Con Ed immediately after the burglary. Con Ed assured him that redemption of the stolen bonds would be blocked (via a "Stop Transfer" order) and new ones would be issued to him. However, the Weisingers had to pay $652 (3 percent of the bonds' market value) to cover the cost of a "Lost Securities" surety bond. When Mr. Weisinger protested, he was told:

"Sorry, that's our policy. We have hundreds of stockholders whose certificates are stolen or lost every year. They, too, have to post a surety bond."

Also missing were several savings-bank passbooks. Calls to the banks were made, resulting in assurances that each account would be closed at once and new passbooks issued as soon as the owners signed affidavits.

Mort Weisinger was less fortunate with another stolen treasure: a copy of the first issue of *Superman* comics, worth at least $3,000, which he (a former editor for the publisher) used in his lectures on the comics. When he listed the loss on his income tax return, IRS questioned it. How much did he pay for it? Nothing, his publisher had given him the copy as a souvenir. The loss was completely disallowed.

After a burglary you are wounded but not slain. So take stock of what happened and how it happened. Resolve not to be suckered again. You are joining good company. Although some victims weep and wail, subconsciously awaiting the next invasion, others do something about it. They spend the next few days finding out a few things about locks and alarms, about good security habits. They talk to a locksmith who does more than cut keys and make repairs after a break-in.

The local locksmith is becoming more and more professional. He is no longer satisfied with merely putting a lock on your door that will keep the wind from blowing it open. A good locksmith is the average man's security consultant. If you talk to one who tells you that "burglars don't pick locks," run from him—he is either living with his head in the sand, or he doesn't have a good pick-resistant lock to sell you. Find a locksmith who does and have him install it.

The concept of having a Crime Prevention Officer in a police department originated in Oakland, California, in the 1960s. Robert McDermott, who was a member of New York's Criminal Justice Coordinating Council, was sent to Oakland to evaluate their program. Returning and reporting to New York's mayor, he helped organize and train the city's Crime Prevention Unit. Since then, virtually every police department in the nation has become involved. Many of these police officers are sent to the

Crime Prevention Institute in Louisville, Kentucky, for training. A call to your local police department may well result in their conducting a security survey of your residence.

PRECAUTIONS TO TAKE

In addition, to save yourself potential heartaches and headaches, take the following preventive steps:

► Set aside an hour this weekend to record the serial numbers on your TV sets, radios, typewriter, camera, watches, other items. For any personal property likely to be stolen, even if serialized, use an etching pencil or cut onto the chassis your name, or driver's license number, Social Security number, address, or other identifying mark. Register these with your local police department. Burglars know that criminal receivers may refuse to buy such identifiable articles or pay much less for them.

► Expensive jewelry should be photographed, the weight of stones and color recorded. Most insurance companies require photos and a "charting" of large diamonds before insuring expensive jewels. Photos of jewelry as well as of antiques can be helpful in their recovery.

► Install an intelligently designed security system of locks and—if necessary—alarms. (See Chapter 11.)

► Contact your insurance agent for an inventory sheet and fill it out. If you wait until after a rip-off you may overlook some important property.

► After taking the inventory, thoroughly evaluate your insurance coverage. Art objects, diamonds, antiques, anything of substantial value should be reappraised periodically by a professional because their value may increase over the years. You may need a more substantial policy. Consider taking out a floater on your policy for treasures such as a coin collection; a floater takes the covered item out of your regular homeowner's policy, so you need less total coverage than you would without it. *Note:* While a hidden wall safe helps convince a reluctant insurance company to cover you, it will probably not reduce the premium. An effective alarm system may.

► When you buy jewelry, have the jeweler give you a written

description of your purchase; if it's diamonds, the color, cut, and carat weight, and a chart of the flaws (carbon spot, fissure, etc.).

► Buying a fur, get a written description from the furrier. Inside a pocket sew a little tag bearing your name and address; the fence may overlook it. Have your initials in the lining, even though the criminal receiver will tear out the lining; the burglar might be grabbed as he's escaping in the street.

Chapter 11 For peace of mind: Mechanical security measures

HOW FAR should you go to make your home safe from invaders? Do you know which security devices are most effectual, which are practically useless?

With the almost limitless range of antiburglary hardware now obtainable it is possible to overdo security, converting your home into a fortress that is both uninhabitable and dangerous—in the event of fire, you may not be able to get out in time. The solution is to know what's worthwhile and to be selective. As we have noted, people who have just been the victims of a break-in often panic, rushing out to spend a lot of money in the wrong places on locks, alarms, chains, and other hardware that won't really solve their security problems.

In the following pages we will describe a wide variety of anti-invasion items that *are* effective. Make your selections from among them judiciously. Installing *all* we mention as effective would cost a small fortune and probably isn't necessary. Select *appropriate* protective measures suited to your particular needs.

Survey your residence, office, or commercial establishment now for security weaknesses based on what you've read thus far. Start choosing, buying, and installing the hardware you are immediately convinced is most essential. *Then* think like a burglar. If you were locked out, how would *you* try to get in? Enlist members of your family to join you in the "game." Where might your home or apartment be vulnerable?

Don't procrastinate, postponing your survey until you hear that burglars are in the neighborhood. You may be the first victim, and that crime may turn into murder or rape. Make your selection according to real requirements; probably you will not need a lock, grating, or alarm at every possible entrance to your residence.

Having taken the proper mechanical measures for your protection and observed the precautions we have recommended, you will have developed your own M.O. for security.

Now let's scrutinize the various kinds of devices obtainable for what we can call a Safe Life.

What's best about locks

Most burglars, it's been established, prefer to work through doors. And for their own safety they don't want to work too hard, too noisily, or too long. They go for homes or apartments that can be entered without the need for obvious and bulky tools. A door lock that's hard to force or pick will send the prowling thief in search of an easier target. So your primary concern is to mount a lock that will make the use of a celluloid strip impossible; that will resist forcible entry with a jimmy; and that will protect the cylinder against forcible removal. For your personal safety, the lock you choose must be fitted with the most pick-resistant cylinder you can buy.

Here's how a typical door lock works. The lock immobilizes the door by means of a bolt connecting the door to the surrounding wall (doorframe). Exceptions are the "police" lock that connects the door to the floor, and the multilocking types that fasten the door to both the frame and the floor.

A cylinder is the replaceable part set into a lock that accepts the key. Use of the key to turn the cylinder causes the bolt to engage or disengage. (Cylinders are picked; locks are forced or 'loided.) The conventional lock cylinder consists of a row of five or six pairs of small pins or tumblers. These are arranged so that the notches cut in your key align them into a precise position (called the "shear line"), enabling the cylinder's core to turn, and thus to move the bolt.

Some lock cylinders are variations on this basic theme: one has twelve pins set in three rows of four pins each, above and on each side of the key slot; others have the pins set both on top

Cylinder guard ring.

and bottom or in a circle. These deviations from the standard are designed to thwart the pick man and complicate unauthorized key duplication. Some are effective, but most succeed only in providing a false sense of security.

A *cylinder guard plate* is a heavy metal piece covering all but the key slot on the outside of the door. It is usually bolted through the door. A *cylinder guard ring* or security collar is aesthetically more pleasing and, if made of steel, does the same job, preventing the forcible removal of the cylinder (see illustration).

How many locks should you have on your door? One store owner put *five* locks on his door, bolting only three of them. His mistaken idea was that if a burglar tried to pick his way in he would be locking some of the bolts he thought he was unlocking. The professional pick man knows when he is unlocking or locking a bolt.

You don't have to resort to Machiavellian schemes to safeguard your home or business. One good "matching set" of lock, cylinder guard, and cylinder will do more to frustrate a burglar than buying and mounting ten pounds of useless hardware.

The following locks are universally available; to assist you, we will spell out the relative merits of each:

INTERLOCKING DEAD BOLTS This is a "rim" lock. It is surface-mounted on the inner side of the door, with the entire lock visible from the inside, and requires use of a "rim cylinder" to activate it. The bolt of this lock moves *vertically* in and out of a strike that is fitted on the frame of the door (see illustration).

These locks, often referred to as drop-bolt or jimmy-proof, have for many years been the most widely used auxiliary locks in the country. They do not offer sure-fire protection against jimmying, but the best of them do offer high resistance to it and to entries where the frame of the door may be spread, allowing a short horizontal bolt to be withdrawn.

An interlocking dead bolt is fairly easy to install on wood or metal doors. It's an absolute necessity for double (or "French") doors. Care must be taken in installing to assure that the vertical bolt slips easily through the vertical holes in the strike. If your door is recessed deep in its frame, the locksmith can use "raising brackets" so that it won't be necessary to cut away a lot of wood (weakening the frame) in order to have the strike align with the lock mounted on the door.

The dead-bolt lock, popular in Northeast urban areas under the Segal brand name, has not enjoyed nationwide acceptance by householders because it is unattractive. Now, however, it is available in a variety of metal finishes to suit virtually any decor. Ask for Emhart's "InterGrip" (see photo, single cylinder, page 220).

Interlocking dead bolt. A Segal "jimmy-proof" rim lock.

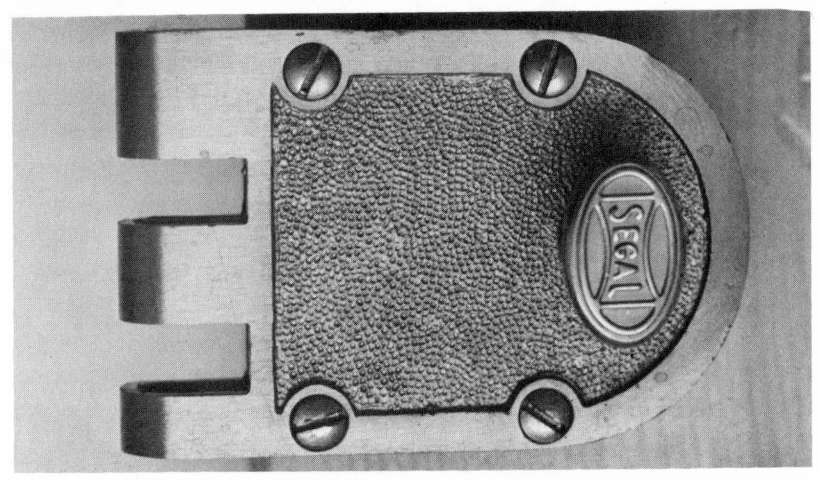

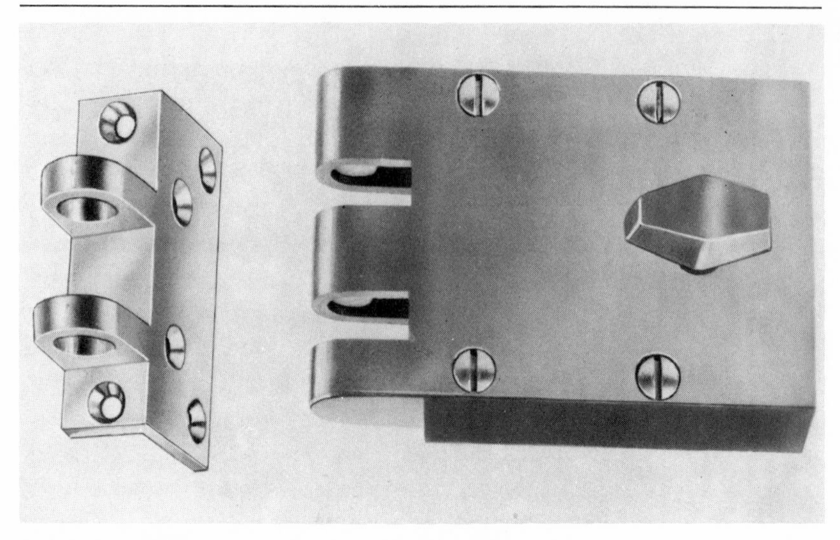

Emhart InterGrip lock: an auxiliary for doors without glass.

In choosing an interlocking bolt lock of any brand, be sure that it's made of solid brass or bronze, and avoid any made of die-cast material.

If your door is fitted with glass panes, can be reached from a nearby window, or its wood panels are lightly constructed, use a double-cylinder interlocking dead bolt. Then an intruder reaching through a break in the glass or wood panel won't be able to merely turn the inside knob and enter (see photo, double cylinder). This surface-mounted lock requiring key use inside will also call for one-way screws to secure the lock to the door. (Reaching through and coming up against your inside cylinder, the burglar will not then be able to use a screwdriver to take the lock off the door.)

Although most drop-bolt locks have some built-in protection against successful entry if the cylinder is pulled or wrenched out of the lock, we recommend that you use a steel cylinder guard ring in addition.

Rim dead-bolt locks are available with a "lockout" feature, used when you are inside. By pressing a button on the inner side you prevent the lock from being actuated from the outside with *any* key, even a duplicate of your own. When the lock can't be opened with a key, obviously a picklock will be equally impossible. But this feature offers no protection against picking when you are out of your residence, and lockouts should never be used on rooms occupied by a seriously ill person or those living alone.

Protection against the quick and quiet pick man is assured only by use of a cylinder he can't beat—a cylinder compatible with the lock in which it is used.

MORTISE LOCKS These are typically installed in apartment doors and referred to as "house" locks. Most are equipped with both a dead bolt and a convenience spring latch (or snap lock). (See photo on page 222.) The *latch* can be set so that it opens whenever you turn the inside or outside knob. The *dead-bolt* part requires a key to lock the door from outside.

A mortise lock is set into a deep recess in the edge of the door, the wood or metal cut out (mortised) to accommodate the lock case. The lock itself is not visible; from inside all you see is a thumb turnpiece, used to open or lock the dead bolt from within. If you lock it with a key when you go out, the dead bolt will be effective in thwarting a 'loid.

You have some security because the locking mechanism lies inside the tough wood or metal enclosure in the door. A burglar could pound off the doorknob, but the door would not open. How-

Emhart InterGrip rim lock, double cylinder: for doors with glass.

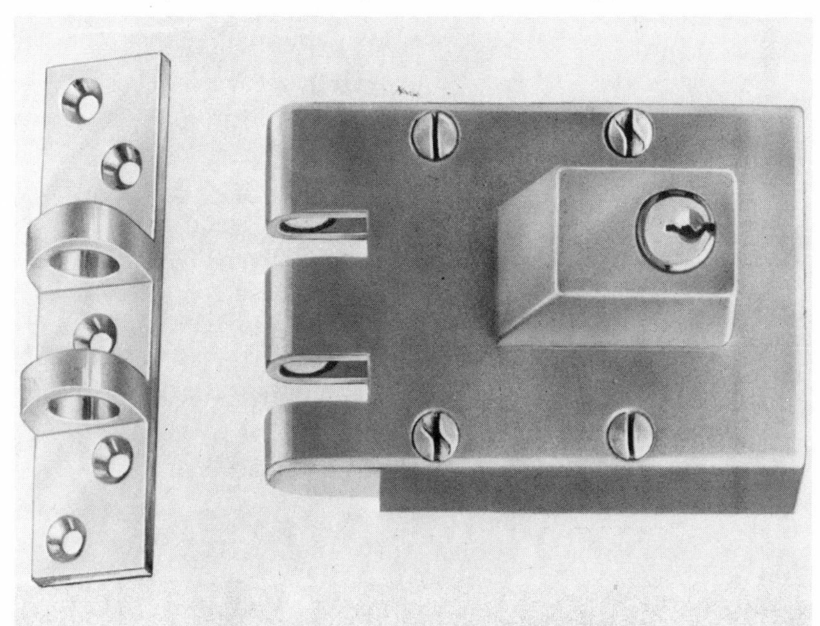

ever, if you neglect to use your key and merely slam the door, you're relying solely on the spring latch—and the lock is vulnerable to a 'loid.

Rarely will an apartment dweller or homeowner add a mortise lock to a door, as it is the most difficult to mount. In apartment, office, and hotel construction, doors are generally delivered complete with a cavity to accept the mortise lock. But if you do want to add one, a mortise lock can be purchased that will be key-operated from within as well as from the outside.

Mortised locks are available with double cylinders, key-operated from both sides. Even in an apartment having no glass panes in the door, a lock that requires use of a key from inside can prevent a thief from carrying bulky or cumbersome articles out of your residence.

CYLINDRICAL DEAD-BOLT LOCKS When this dead bolt is mounted, nothing of the lock is visible on the inside of the door except a thumb turnpiece; or if it is a double-cylinder lock, you will see only the cylinder.

The vertical bolt on early versions of these locks extended only ⅝ of an inch when locked. Later models, better for security, have a full one-inch bolt fitted with a hardened revolving-steel

Typical mortise dead bolt lock.

insert to prevent the bolt from being forced or sawed through. The added bolt length safeguards against spreading of the door-frame and jimmying. Protection against wrenching or prying the cylinder is provided by steel collars that are part of the lock.

Cylindrical dead bolts are easy to install, and are rapidly becoming the most popular type of auxiliary lock. The best of them, when mounted on wood doors, give the highest degree of protection against all forms of forcible entry, a claim verified by police tests. Available in "double-cylinder" for doors vulnerable because of their glass panes, they are versatile and very effective—in single-door applications. We do not recommend them for double-door use; on these, as pointed out, the *interlocking dead bolt* is best.

If you use a highly pick-resistant cylinder with this lock, you will have touched all the bases for single-door use. If you are building a new house, you might install them on all perimeter doors. Use them in conjunction with "standard duty" key-in-the-knob lock sets—to prevent easy invasion of privacy during the day—and use a combination escutcheon plate (a decorative front piece) to house the two locks on each door. Have all the cylindrical dead bolts and the cylindrical lock sets (key-in-knob) keyed alike so that the one key on your ring will open all of the locks—even if the cylinders are all high security; that is, pick-, drill-, and shim-resistant in the cylindrical deadlocks and conventional in the key-in-knobs.

In any double-cylinder lock installed, leave the key right in the lock or hanging on a hook near the door while you sleep so you can get out in a hurry in the event of fire.

Unless you have very unusual house or apartment doors, relying on either the rim or cylindrical auxiliary locks described will greatly improve your odds against being burglarized.

If you are building a new house, instead of the cylindrical or rim, you may opt for a mortise lock to be installed for you. Be sure it has a bolt with a one-inch throw (bolt extends one full inch from the face of the lock) and that the bolt is reinforced with a hardened-steel insert. Use the companion high-security cylinder to protect against picking or other cylinder attacks and it will be necessary to protect the cylinder in the mortise lock from forcible removal by adding a steel cylinder guard ring.

In selecting a mortise lock for the builder to install, be sure it's equipped with an armored face plate. This covers the set screws holding the lock's cylinder in place.

A lock with an interesting configuration probably best described as a cylindrical lock with a dead-bolt function—but also having characteristics of a mortise—is the good-looking Schlage "G" series (see photo). The lock mechanism is protected from drilling by a metal plate hidden under an escutcheon; this model also has a recessed cylinder to make it hard for thieves to get at the cylinder. The lock cylinder itself, however, leaves a great deal to be desired. So despite its one-inch throw bolt the companion cylinder invites surreptitious entry.

The locks discussed thus far will serve for the vast majority of doors. Where door conditions are unusual, others may be essential:

THE "POLICE" LOCK The so-called police lock is designed primarily for use where the door is strong but the frame is weak. In a unique feature, the lock bolt makes no contact with the doorframe. It depends for locking on a heavy steel strut (a brace bar) that is propped up at an angle connecting the door to the floor (see sketch).

Any force used against the door is transmitted to the floor, making it highly unlikely to break the door open by normal brute force. While unsightly, the "police" lock is effective against forcible entry by kicking in or jimmying. There is a way of defeating these locks—but we won't reveal it here, since there's no

Schlage "G" series lock, a cylindrical/mortise lock with a dead-bolt function. It's protected from drilling by a hidden metal plate.

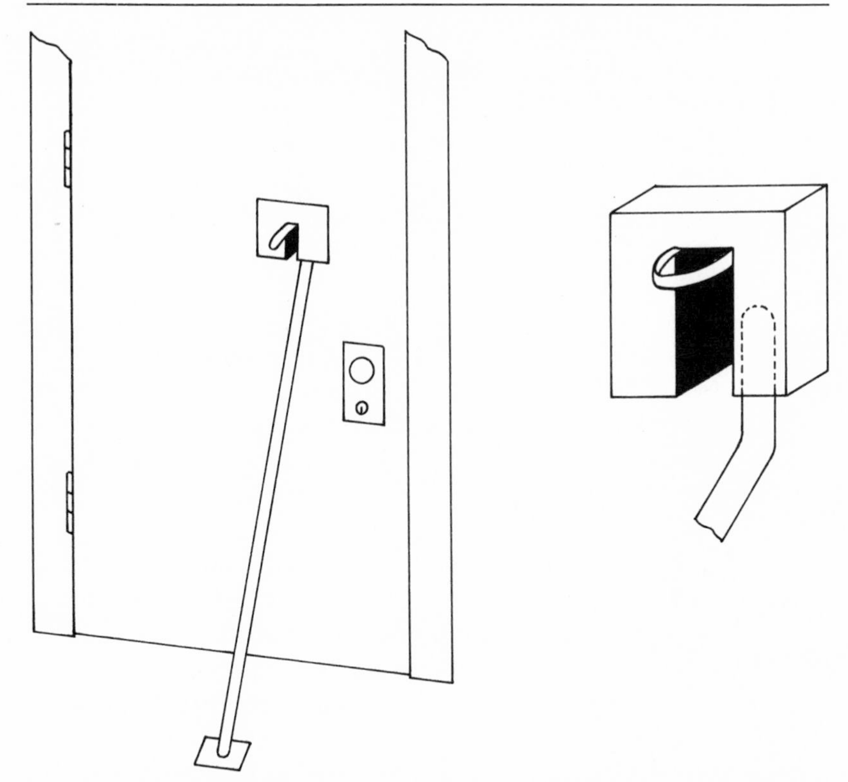

The "police lock" made by Fox Lock Company and Magic Eye Lock Company.

other type of lock for the weak-frame-strong-door situation. "Police" locks are made by the Fox Lock Company and Magic Eye and can be used with pick-resistant cylinders.

How did the name for this lock originate? There are two theories. Police on patrol often find an "open door" in business establishments where the proprietor has failed to secure it. They "lock" the door by nailing a piece of wood to the floor about eighteen inches inside the door, then propping a two-by-four against the door and slowly closing it; as the propped piece falls, it effectively (and rather permanently) barricades the door. The lock may have taken its name from that police practice. Another theory holds that so many cops broke their shoulders against barricades like this that the lock was named for the police.

In the late 1960s and into the 1970s many young people took up residence in commercial buildings and warehouse lofts never intended for the purpose. Most doors at these places open outward, exposing the hinges to attack. Occasionally this condition is encountered in private homes.

In a private residence we recommend two courses of action:

Change the hinges to nonremovable pin types or immobilize the pins in your present hinge by one of the following methods:

Weld the pin to the hinge.

Drill a small hole through the hinge body and hinge pin, and insert a small nail or pin into the hole, flush with the surface. With the door fully open, insert two large screws into the door edge near the hinge, one screw near the top of the door and the other near the bottom. Leave the heads of the screws exposed about half an inch. Then drill two holes in the jamb so that when the door is closed the screw heads fit into the recess. Now, even if the hinges are removed the door cannot be removed.

In a commercial building, on doors opening out to corridors, in addition to securing the hinge pins we suggest you use a Fox Lock (see illustration). Here the cylinder is mounted in the center of the door protected by a heavy steel plate. Rotating the key in the cylinder causes the bolts, which engage on *both* sides of the door, to be withdrawn. Chances of cylinder removal or jimmying are remote. Another excellent lock for this situation is Sargent & Greenleaf's "Barricador."

A problem for many householders is the split or "Dutch" door that have glass panes in the upper half. If this is your situation we suggest you install two cylindrical dead bolts, one in the top portion, the other below, both double-cylinder, and the two keyed alike for convenience. If the doors and frames are solidly constructed and there is no glass in or near the door, put on two shiny decorative heavy-duty brass sliding barrel bolts, one for the top half of the door and the other for the bottom. These are seldom the only perimeter doors, so secure the bolts and exit through another door when leaving your home.

PUSH-BUTTON COMBINATION LOCK No key is required for the unique type push-button lock (see illustration on page 228). It has no keyhole, and it can't be picked. But its advantages are limited, so it should be used only as an adjunct to your security. The combination lock is handy if you want to give the last two of four digits to a domestic, a workman, or anyone else authorized to enter your residence while you are out but to whom you don't want to hand over your keys.

To take advantage of this feature you'll have to leave your cylindrical or interlocking bolt lock unlocked. On leaving, you lock your door and then press the first two digits; the domestic or workman has the remaining two, but not the entire combination, limiting his entry to this occasion only.

Note: This lock is not for prime protection. While an expert can't feel out the combination on either the lock having parallel banks of buttons or the one with the buttons set in a pentagon shape, given some time he can dope out the sequence code.

The Fox Lock. Here the cylinder is mounted in the center of the door under a heavy steel plate. Forcible entry is extremely unlikely.

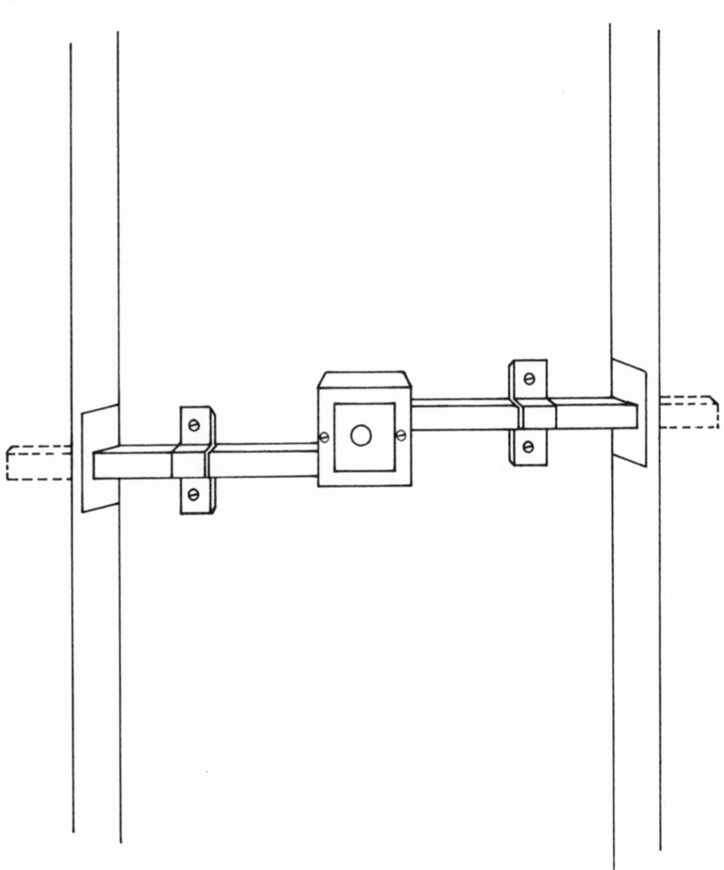

Because the manufacturers instruct users that they may press two buttons simultaneously in any code they select, people invariably will do just that. Starting with that knowledge and then working patterns rather than numbers, it won't take a professional burglar too long to determine the combination. Once that combination is known, a thief can visit you at his leisure and you will never prove a burglary to collect on your insurance.

True, you can change these combinations whenever you like but we believe you won't do that very often. The push-button lock is very convenient for some interior doors (like the closet where you house your sound equipment) and excellent for lavatories in office buildings, but it should not be heavily depended upon in the models economically feasible for homeowners. There are some very sophisticated types available, electronically equipped to set off an alarm if the wrong digits are worked. These may be appropriate for people concerned with security over computer rooms and the like.

A somewhat similar *knob-combination* lock, which also can't

Push-button combination door lock.

The 3-M Lock Alarm.

be picked, is more vulnerable. A burglar familiar with the lock can simply rotate the doorknob until he "feels" the lock make the various adjustments for opening. Cheap locks of this kind are made so crudely that even an amateur can find the combination by experimenting.

3M LOCK-ALARM Manufactured by the Minnesota Mining Company, this is a preentry local alarm device (see illustration). It causes an alarm to go off when any attempt is made to pick or jimmy it, to pull the cylinder or force the attached chain lock. The lock itself is fairly sturdy and when properly mounted will resist force.

To set the alarm, simply lock your door as usual. We believe your security will be better served if you take advantage of 3M's offer, as an option, of a Medeco cylinder for use with the unit. In our opinion the alarm on this is not loud enough to scare a determined thief away from your residence. The alarm actuated is shut off by pressing a button on the lock case.

MULTIPLE-BOLT LOCKS Multiple-bolt locks are fairly new to the residential security scene. These utilize a single cylinder to operate bolts that engage into the doorframe at several points. A French model, Fichet, sets three bolts—one on top, another at the bottom, and a third that moves horizontally into the frame. An expensive and difficult lock to install, it also leaves a great deal to be desired aesthetically. The Fichet company also markets a surface-mounted (rim) horizontal bolt lock. In tests, the mounting screws were found to be too weak and easily forced, while the flat strike plate was rather weak.

KEY-IN-KNOB LOCKS Do *not* rely on the key-in-knob lock (see illustration). It may be used only to give you a modicum of privacy during the day. But even for this function, we recommend you avoid the "light duty" models and select only "standard" or "heavy duty" lock sets. Key them alike with your auxiliary lock, which is providing your real security.

PADLOCKS In many situations, only a padlock will serve to safeguard property: on gates, sheds, basement windows, and slipped into the track mounted inside your garage for the overhead door. The security padlocks offered range from the ridiculous to the sublime. Look for a hardened steel shackle and solid brass or steel body. You may find a combination type convenient but in general avoid those with rotary spin dials. Most of these have coded numbers stamped on the body, and the codes are published—giving a thief the combination! Prefer a combination padlock (or combination luggage lock) that permits you to set your own combination (see photo).

Key-in-knob lock—a spring-latch, standard-duty type. Any force on the knob or latch with a wrench or jimmy easily defeats it.

A combination padlock. User sets his own combination and can change it at will. The same type is available in luggage locks.

Happily, well-made padlocks are also obtainable with removable cores. With these you can have your locksmith key them the same as your front-door lock; then you can use the same key for both.

TRAVEL LOCKS Staying at a hotel, you normally have little control over the security of your door. We recommend packing a small, inexpensive travel lock in your suitcase. This key-operated lockout device can be slipped onto the inside of just about any door to prevent it from being opened from outside, even with a key. It has a small narrow hook on a piece of steel that fits into the strike plate of the commonly used (and largely ineffectual) key-in-knob lock. Simply sliding another steel part over this hooked member will insure you an uninterrupted night's rest. Made by Yale, the locks are an excellent investment at under $10.

SELECTING A CYLINDER AND THE LOCK FOR IT The most pick-resistant cylinder today is Emhart's High Security, manufactured by the Hardware Division of Emhart Industries. The cylinder (see illustration on page 232) makes use of a totally unique set of interlocking pins and followers (drivers and tumblers), and the cylinder's face is protected from drilling attacks.

A great advantage to the householder lies in the fact that the key to operate the High Security cylinder will also operate in

Emhart's conventional cylinder. Thus, if you need a double-cylinder lock on a door because the inside of the lock could be reached through a pane of glass, you can obtain one of Emhart's locks with a High Security cylinder on the outside and a conventional cylinder inside. The High Security key will operate on both sides of the lock. This saves you the inconvenience of having to use two different keys to operate one lock, and it's less expensive than buying two pick-resistant cylinders.

Expert locksmiths have long tried in vain to pick the Emhart. Listed by Underwriters Laboratories, which indicates the lock met standards for burglary resistance, it's used by federal agencies and is available at your local locksmith. It can be a replacement for the cylinder in your present lock, or may be used with Emhart's InterGrip ("jimmyproof") or their cylindrical deadlock.

The following cylinders (see illustration) may be used to advantage if your present locks meet the criteria we have set earlier:

Medeco: In our opinion, this lock offers excellent security

The Emhart High Security Cylinder. This cutaway view shows the unique interlocking of the top and bottom pins to prevent picking, use of manipulation keys, and defeat by the impression method. This feature, and the shield and armor rods to prevent drilling, make for an excellent security investment.

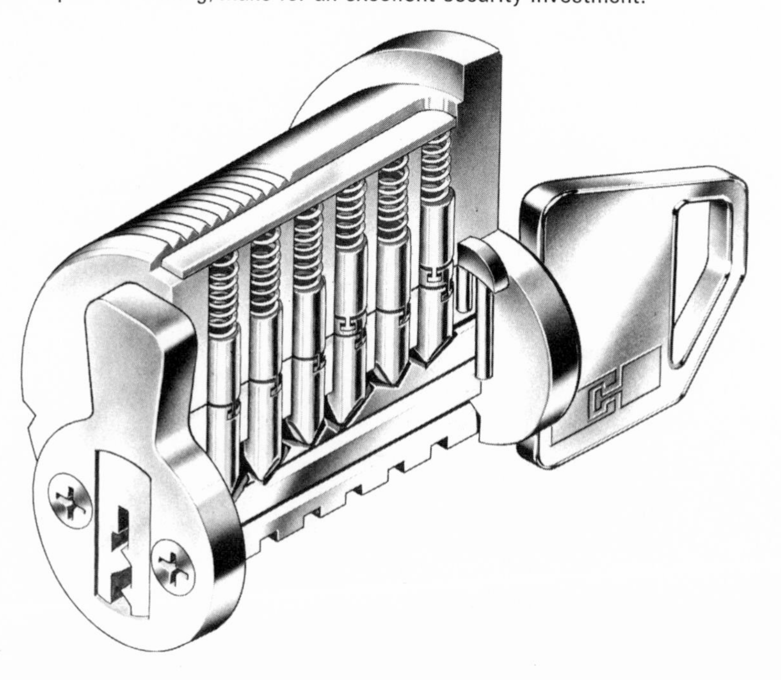

Faces of six pick-resistant cylinders. **Left, top**: Medeco; **left, bottom**: Ace; **middle, top**: Emhart; **middle bottom**: Sargent Keso; **right, top**: Fichet; **right, bottom**: Magic Lock. Of the cylinders shown, only Emhart and Medeco meet our criteria for high security.

against picking. The Medeco, having no companion lock of its own, may be used to replace the conventional cylinder in your lock. The sophisticated tools that are available to defeat this cylinder we do not regard as a pertinent threat to household security.

Fichet: These may be used as replacement cylinder in American locks, but we do not recommend their Multi-Point or rim locks.

Keso: This lock is made by the Sargent Lock Co. Despite stories you may have heard of exotic methods of defeat, the Sargent Keso may be accurately rated as a pick-resistant cylinder. *Note:* If you want a duplicate key made for a member of your family, you may have to wait a while because your key must go to the factory.

The following cylinders are better than "conventional," but still "iffy":

Magnetics: Magic Lock, with a disk-shaped key. Duplication is a problem for the householder, as you must order from the manufacturer. They offer a standard-duty key-in-knob lock set but no interlocking bolt or cylindrical deadlock.

Ace: A round key. Beware. Don't use it in an exterior alarm switch or to secure your door or a coin-operated machine. If you do, you are flirting with the burglar who has bought a $16 tool.

Segal-Hines key system is not recommended.

With any cylinder you select, install a good steel cylinder guard ring or guard plate. In our judgment the more attractive guard ring will provide enough protection against a burglar unable to pick the lock who takes the next step of trying to forcibly remove the cylinder. But your use of a good cylindrical deadlock will make a guard ring superfluous.

WHERE TO USE THE LOCKS

Where to install your locks is as important as what kind you choose. Here's a simple guide:

► Where the door is strong but the frame weak—and no glass panels are in close proximity—use the "police" lock.

► On a metal door set in a steel frame your best bet probably is the vertical interlocking bolt, such as Emhart's InterGrip or Segal's Jimmyproof. On French (double) doors in a residence, only the InterGrip type will serve best.

► On wood doors solidly constructed and set in a sturdy frame, either the cylindrical dead bolt or the interlocking bolt rim lock will serve. If there's a glass pane in the door or the door panels are frail, choose a double-cylinder lock and install it with two one-way screws. Make sure that one key for the double-cylinder lock works on both sides of your door.

► On any door, prefer a High Security cylinder to prevent picking, drilling, shimming, or a key being made by impression. (*Impression* is a technique for determining the depth of the cuts in your key by blackening a key blank and filing it until the key works.)

► It's never necessary to have more than one High Security cylinder on a door.

Finally, choose a locksmith who is a member of a professional locksmith association or one recommended by the Better Business Bureau or another consumer organization. As a whole, locksmiths are a wonderful breed. During twenty years in the New York Police Department, McDermott heard of only one who was arrested for burglary in the city, and he had lost his license before his arrest. Still, you may feel a little more comfortable if your High Security lock and cylinder come in a factory-sealed package.

DOOR CHAIN BOLTS AND LOCKS Any door in your residence where you may need to talk to salesmen or other strangers should be equipped with a chain bolt. A battery-operated interviewer at the door can be installed; but for most purposes a good sturdy chain and peephole will suffice.

Do *not* depend on a chain bolt—no matter how it's installed—to protect against burglary. It is not a safeguard while you're out even though you've set it with a key (for a key-controlled chain bolt) or at night while you're asleep, believing no one can get through the chain without noise and force. (Remember the Susan Clarke rape case.)

Yes, a chain bolt has its purpose—and limitations. Whether or

A Central Station alarm company's monitoring system.

not it's key-controlled, use it the way it is effective, to interview the fellow you're wary of.

Your choice of alarm systems

The greatest single deterrent to burglary is alarm equipment that thieves have learned to respect. Don't make the mistake of thinking that *any* alarm setup is better than none. A poor one may give you a false sense of security, lulling you into overlooking other forms of protection. You may leave valuables carelessly around, fail to record serial numbers, neglect to mount pick-resistant locks.

Good alarm devices and systems have proven their worth. They are versatile alarms you can live with, dependable, and simple to operate. And these virtues are known to professional burglars.

CENTRAL STATION ALARM With a central station alarm (see illustration), a central station company installs sensing equipment in your residence, office, or business establishment designed to detect the presence of an intruder. Leased telephone lines transmit a *silent* signal to a central office manned around the clock by trained operators. When you subscribe to such a service you decide what response should be made when an alarm goes off. You can have uniformed armed guards rush over; response by guards and police; or response by police only. Banks, jewelers, and furriers—as well as private residences—depend on central service systems such as those offered by the Holmes Protection company, Burns, and American District Telegraph (ADT).

The equipment installed on the premises requires you to use a coded signal to the monitoring central office when opening your door. Any deviation from the code results in response by guards or police. The homeowner or apartment dweller opting for this ultimate in alarm protection will rarely ask for the most advanced "line security" techniques. But commercial high-risk premises must insist on the best available in their area.

The first really sophisticated penetration of a line-security system occurred in California in 1965. Three years later, the disastrous safe-deposit-box burglary in the Manufacturers Hanover branch in Forest Hills, New York, sent shock waves through the banking community. Thieves, prying open 178 safe-deposit boxes, made off with loot estimated at from $3 million to $7 million (see photo). An FBI informant later fixed the score

Scene at a branch office of Manufacturer's Hanover after burglars broke in and rifled 178 safe-deposit boxes.

at $12 million, almost all of it in cash, jewelry, and bearer bonds.

A block away from the bank building, the burglars had compromised the central office alarm telephone lines. They had rented an apartment on the second floor of a nearby building. Below the level of the apartment windows, telephone cable housing the bank's alarm lines was strung, tacked against the side of the apartment building. Working from this point the burglars utilized electronic equipment to defeat the bank's "basic" alarm system.

Of course, line-security capability has been much improved since then. Available now to the high-risk business community in many urban areas are computer-monitored systems. With these, the computer sends out random electronic pulses over the telephone line and then interprets the pulsing signal back to it.

A responsible protection company will install this kind of system only if the subscriber has heeded their recommendations about physical security. The computers they install are costly; and when that decal goes on, announcing to all the world that the door of the businessman or bank is protected by the company, their reputation is at stake. They know all about picklocks and fast entries. They fear the use of ordinary cylinders on locks, because they need time to respond. If the burglar has to resort to force to get in, the responding agents have that much more of a chance to get to the scene while the intruders are still at work.

About six weeks after the Forest Hills bank was gutted, Manufacturers Hanover again burst into the headlines when another of their branches—at Lexington Avenue and Fifty-first Street in Manhattan—was ripped off. A different central office alarm company served this location but the result was the same—a successful burglary through defeat of the alarm system.

Again there was only the basic alarm system, with none of the available "line security" features to challenge the electronic skills of the safecrackers. In this crime the crooks were able to work on the alarm system right at the control box inside the bank. Only the vaults were protected; with none of the perimeter doors wired for intrusion, the burglars picked the conventional cylinder on the bank's door, entered, locked the door behind them, and unpacked their tools at their leisure.

CENTRAL OFFICE ALARM PROTECTION FOR HOMES AND APARTMENTS Certain central-office monitoring systems are economically suited to apartment tenants and homeowners. In cities where a timely response by armed service guards and/or police is feasible, these systems offer a high level of protection. A central office using leased direct telephone lines constantly monitors the residence setup. In many of these installations you'll have to go through "opening and closing" routines, signaling the company when you enter or leave your dwelling.

There are systems for home use that are equipped with time-delay features. With these, you have twenty or thirty seconds after you come in through a protected door before you must abort the alarm by using a special key in the control instrument.

One of these systems is the Apartment Intrusion Detector (AID), popular in New York City. There the Holmes Protection company can connect the lines to their Computect System, which monitors faithfully by computer. But even an AID system won't stop an intruder if you make a habit of leaving your control key in the instrument located inside your apartment.

One burglary of a luxury Fifth Avenue apartment happened just that way. The victim was the wife of an influential newspaper publisher. After the burglar picked through the ordinary apartment-door lock in seconds, he walked directly to the alarm control box and aborted the alarm, using the key left in it for his convenience.

Generally these keyed shunt switches are operated by a pick-resistant cylinder, so that if a burglar makes a break into the residence through a protected door or window and he triggers the time-delay sequence, he will not have enough minutes to successfully use picks in the lock to shut off the system.

A very desirable feature of AID and other systems available from Burns and ADT is a panic button which can be termed a holdup or panic signal transmitter. Usually mounted near the entrance door, this remote control unit—when actuated by finger pressure—will cause an alarm to be received at the protective service's control office nearest you and initiate emergency response. Since no bell or other sounding device is heard in the home, it doesn't tip off the intruder. The button can also be used on a bedroom night table to summon help in the event of a sudden illness by a person living alone.

Instead of a silent signal, you can opt for an audible device, such as a bell, to frighten an intruder into fleeing.

Approximate cost of this type of system is $225 for initial installation, plus $25 a month for the leased phone line, maintenance, and twenty-four-hour monitoring. Service charges are reduced if a number of tenants in your building subscribe.

Private homes or apartments desiring protection at more than five points (windows, doors, etc.) will not be able to utilize the AID system, but similar features are available in more extensive installations.

In outlying areas where there may be no central-office monitoring facility close enough for good response, many of these same companies (protecting banks and jewelry stores in the cities) can still offer effective protection by making a "police connect." If your local government permits police involvement in monitoring alarms from private residences, we suggest you investigate the economics of this option.

The police departments that do accept alarm lines into their quarters will be very particular about the company doing the job for you and the type of system installed. They have learned from harsh experience to be choosy; the false-alarm rate from a poorly installed system or one using inferior equipment makes police involvement prohibitive.

But once a good system like Holmes works for you, so do the police. They monitor and will respond to any signals received. These police connects from private homes invariably make use of a keyed shunt switch for on-off control. Make sure yours is mounted inside the perimeter of protection and that you use a time delay.

PROPRIETARY SYSTEMS These are installed throughout an entire building or group of buildings in close proximity to one another. They are monitored from a location in the immediate area, often from within the buildings themselves. For example, at an apartment building all tenant doors may be wired to detect unlawful entry, with every installation monitored from a console located within that building. In most cases, these systems call for the building's guard employees to respond to the (usually silent) alarms. Generally a proprietary system does not make use of or need sophisticated supervised circuitry, although certain installations are notable exceptions.

Some proprietary systems are excellent, others practically worthless. Too many variables at each installation must be evaluated. There are companies experienced in proprietary system installation who can be relied upon to do a creditable job. Among them are Mosler's Electronic Systems Division and the Holmes Protection company. Much depends, of course, on the quality and training of the responding guard personnel.

LOCAL ALARMS When such an alarm is activated by an intruder, a bell or siren goes off in the immediate vicinity of the invasion. The deterrent, then, is the clamor. There's no provision, however, for automatic police and/or company guard personnel response.

This type of protection is not recommended for premises (residential or commercial) that might be classified as high-risk situations. There, only central-station protection will be effective. Local alarms will prove inadequate because even the best are vulnerable to sophisticated attack, and the worst can be fouled up easily.

In commercial premises not particularly attractive to the thief, a good local alarm may be used to advantage. In a residence, the assurance that you get with a central station system—knowing that help is on the way—can't be ignored.

In your local system, a panic button mounted near the entrance door may serve to good advantage—if properly used. In speaking with a caller at your door (your chain bolt engaged, of course), if he should try to force his way in, pressing the panic button will cause the alarm to sound, and he will almost surely depart. However, if you should sound the alarm when a person you fear is already in the apartment it may drive him to make a reprisal. If you were linked to a central-station unit, the panic button would summon aid silently.

Local alarm companies are too numerous to list; most central-office alarm companies install "locals" as well.

For the average citizen, the local alarm may be the device to use. If your apartment or home has more than five vulnerable points to be protected, this would preclude AID; and the more extensive wiring needed for other systems may be beyond your budget. If you are in the enviable position of having a neighbor who is usually home and who would call the police when your alarm goes off, you may opt for a good local system. For most people the problem is determining just what makes for a "good" system. We suggest you look for the following characteristics and features in your local alarm setup:

The equipment to be used and its installation should meet the standards required by Underwriters Laboratories (UL). Remember, these represent the *minimum* standards acceptable. You may, in seeking the proper alarm for your needs, find that two companies are listed with UL. This doesn't mean that both perform equally well; learn something of each organization's track record. In the field of security reputation is paramount. Find out from the firm:

How long have they been in business?

How extensive is their operation?

What kinds of premises do they list among those they service?

While the devices they install are approved by UL, the quality of their installations should be determined through references.

How many people do they employ in their service department?

Do they service and maintain the system or do they merely sell and install, then leave you on your own?

TELEPHONE DIALERS Telephone dialer alarm systems, currently popular, are activated by a variety of triggers. On intrusion a prerecorded message is sent over a telephone line to whomever the subscriber chooses. The message may say: "This is John Jones at [his address]. A burglary is in progress here."

Most of the systems are programmed to call the police first (911 or other police emergency number) although some departments have prohibited this practice due to a high false-alarm rate. Then a call may be made to other pretaped numbers—the owner of a store or a neighbor, for example. In remote locations where an audible system might not be heard, where police connection is not possible, or where central-station service isn't available, one answer is the "dialer." You subscribe to a system provided and serviced by one of several telephone-dialer installation companies such as Holmes and other reputable central office firms. Where this system is used, we suggest that the second or third person notified should not take it for granted that police have been summoned. They, too, should call for police.

Unfortunately, with the present state of the art, there are several serious disadvantages in these systems. The biggest drawback is that cutting of the phone line negates your protection.

Also, the system is easily defeated if a thief has your phone number. This is true even with dialers designed to cut off incoming calls after three or four rings. A burglar with your phone number will come to your door, pick your lock, and unlock the bolt without opening the door. Then he will signal to a friend waiting in a car who will dash to the nearest public phone and call your number. The moment the phone starts to ring, the thief at the door opens it and slips inside. The phone-dialer gadget will then do exactly what the manufacturer has claimed—it will disconnect the incoming call, restoring the alarm to the "alert" status. But it's too late: The burglar is already at work in your bedrooms.

In cities like New York, police report a very high false-alarm or call-not-answerable rate. Some cities have sought legislation to make illegal a direct call to police by these "robot" alarms. An intermediary then has to be used. This has resulted in some companies setting up an "answering service" to monitor, falsely terming this a central-office alarm service.

Taped messages begin when the connection is made with the number called. They are not voice-activated. If the call is put on standby, police or another person called may get only the tailend of the transmission, even if you program the tape to repeat three times.

Bearing in mind these drawbacks, look into dialer alarm service if that seems best for you in your area. Ask questions, though, based on what you've just read.

SPACE-PROTECTIVE DEVICES Here's an alternative to wires, door contacts, taped windows, and other expensive eyesores.

Space-protective (or volumetric) units consist of a sensor installed within a home, apartment, or business establishment to detect the *presence* of an intruder. Usually activation of these triggers doesn't depend on the opening of a door or window. Rather, it "senses" a change within a specific area, indicating unauthorized intrusion. Such sensing equipment is frequently used in conjunction with excellent central-station and proprietary systems. It can also activate good local alarms.

Let's look at those *portable* units being advertised for use in your home. Attractively packaged, they don't require the running of wires or the attachment of contacts on doors and windows. Installation consists of plugging in a line cord—so you may even take your alarm along with you on vacation. But, with almost every unit on the market, only one of two types of sensors is used, both somewhat prone to false alarms.

Good-quality space protective sensors properly installed may well answer your problem of too many vulnerable windows to be wired and foiled at considerable expense. You have a choice of several types:

Ultrasonic alarms: This equipment is designed to fill an area with high-frequency sound waves. Motion within the area covered by the wave pattern will cause a distortion of the frequency from that originally transmitted or beamed by the unit. Since the device detects motion, *any* motion in the protected area will trigger an alarm. This, of course, includes the movement of air currents and of pets through the area. In most cases these units will not detect a person moving very slowly through the sensitized zone. And loud noises from outside your "protected" residence may set off an alarm.

Although described as "ultrasonic," these systems generate sound only just within the upper limits of the audio frequency spectrum. For people and animals who can hear the piercing sound, it is intolerable.

Microwave sensors: One common sensor for portable units is microwave. Making use of the radar principle it is used, as is ultrasonic, to detect the *presence* of an intruder in a specific area. Microwave is also a motion detector as radio waves are transmitted and received back by an antenna. When any motion in the area causes a signal to be received back at a low frequency, an alarm condition results.

Unlike ultrasonic, the microwave is unaffected by either air currents or loud noises. It may be triggered, however, by other radio transmissions in the area if these are operating on frequencies similar to that of the alarm unit. Unless properly adjusted, the wave will penetrate walls and cause alarms by detecting motion in adjacent apartments or other places.

The microphone system: Here, rectangular microphones placed in rooms, and tape-recorded sounds that are picked up, get hooked into local telephone lines to a local security service. The sensitive mikes, which look like thermostats, record such sounds as footsteps, a squeaking door, metal scraping on metal, drawers being opened, even heavy breathing. The system is monitored around the clock by service personnel seated before a large console with panels for each rotation that light up when the mikes pick up noise.

If the mike line is cut at any point, the alarm is set off. When a key turns the system either on or off, the occupant must call the local service office almost immediately each time and give an individual code identification. If the procedure is not followed, police are sent to the scene.

This is being used throughout the country in places ranging from apartment houses and mansions to museums, government buildings, Xerox offices, and General Motors plants.

Stamford, Connecticut, where several of these setups are in use, reports not only a dramatic drop in break-in losses in the city schools but also savings of $100,000 in insurance premiums over three years. In a local department store, mikes once picked up grunts, movements, and a hoarse whisper, "Get that one!" followed by the roar of a car engine. The sounds had been relayed to a police dispatcher who informed a radio car that the burglars had left the store. A block from the break-in, a car with two men was stopped. In the rear seat were two color TV sets bearing the store's tags.

In another case two men arrested for burglary claimed they had been inadvertently locked in a store at closing time. But a forty-minute tape played their conversations as they rummaged around the store for watches and cameras. They pleaded guilty. This microphone alarm system is manufactured, installed, and maintained by the Sonitrol Security System of Orlando, Florida.

This type of protection would be "overkill" in a home or apartment.

MAKING YOUR SELECTION Should you buy an alarm system outright or contract to have a system installed and maintained, paying a monthly rental and maintenance fee? We recommend the latter.

Alarm companies that require the equipment to remain their property and who then maintain it will almost invariably do a better job of installation. Most instances of false alarm in any system are caused by poor installation of equipment and improper use of certain sensors.

When you *buy* an alarm system, a considerable investment, you'll have to pay for any necessary repairs, such as cut wires, if it's attacked.

Maybe you've heard of alarms that "rearm" themselves after sounding for a certain period of time. These are generally not perimeter but space-protective devices that detect movement, ambient light levels, sound, etc. They rearm with the cessation of the disturbing factor. A good perimeter system (preferred for most residences) causes alarm on the breaking of a circuit. It will not shut off and rearm itself as long as the electrical upset exists.

The companies with the best reputation in the local-alarm field are those who maintain control over their equipment. You will be saved the problems of alarm repair, periodic changing of batteries, and other headaches. And you will almost certainly have a system less prone to false alarms.

The economical AID system and others like it are designed primarily for apartments; only a limited number of protection points are possible. If your residence requires more vulnerable points to be covered, a local system may be more appealing and serve as well. If you can't use AID but can't afford the more costly central-station services either, consider a local alarm. Central-station companies such as Holmes, Burns, and ADT install and maintain local alarms. If you decide on a local and have it put in by a central-station company, make certain your system will be installed by the same work force that mounts the more sophisticated central system—and that a subcontractor is not involved.

BEFORE YOU BUY To check on the reliability of a service or an equipment dealer, talk to people you know who are using them, and call the Better Business Bureau or other consumer agency in your area.

Avoid a setup with a control panel so complicated that everyone in your household may not understand how it works.

Consider an alarm system that, for a modest added cost, includes sensors for outbreak of fire.

Before subscribing to a central-station alarm system, check to see the number of trained armed personnel at the station ready to answer emergency calls. One man on night duty is not enough unless his job is only to notify police who will respond.

If your system depends on signals to your local police, find out by a personal visit to them how they feel about residence alarm signals.

Whatever system you select, see if you can get a written guarantee for a specified period, and ask about warranties as required by law.

Test your system at every protected point, reset the alarm and test again, before you sign up. The time lapse in the delaying device should be long enough to be handled by all members of your family.

Some alarm systems dependent on the telephone line may not be approved by your phone company. Inquire about it before you go ahead.

DISPLAY YOUR DECAL PROMINENTLY Alarms are a considerable deterrent to thieves. Any alarm company putting in a system—whether it's central office, dialer, or local—will supply you with decals (see illustration). Use them. Don't fall for the bromide, "If I put up a sticker showing I have an alarm the crooks will think I have a fortune in the place—and they'll come to rob me." Don't forget: Burglars are seeking the quick and easy score, and they don't want to get caught.

Decal of Holmes alarm system that user displays.

Safes and chests

A chest is a "money safe" (see illustration). A safe—the kind you're most familiar with—is designed primarily to protect documents from fire damage. It will not resist a determined attack by even an unskilled thief.

The standards for safes and chests manufactured in the United States are, for the most part, established by Underwriters Laboratories, Inc. (UL). Labels are affixed to the units indicating their degree of resistance to various methods of attack and/or the general rating of the equipment. For example, an "E" rated chest must have a door 1½ inches thick, with sides, back, top, and bottom measuring one inch.

Other designations appearing on labels of safes and chests include: *TL* (tool attack, standard tools); *TRTL* (torch and tool attack); and *TXTL* (tool, torch, or high-explosive attack). *TL*, followed by a number, indicates the equipment will resist attack for the stated period of time with standard tools used. For example: *TL 15* means that the unit will resist an attack with standard tools for 15 minutes. *TRTL 30*: resists torch and tool for 30 minutes. *TXTL 60*: resists attack by expert burglars using tools, torches, high-explosives or any combination of these for 60 minutes.

It is only in the last classification (TXTL 60) that the laboratories concern themselves with anything other than the door and front surfaces of the chest (safe). All other classifications seem to assume that the attack must take place on the door. Since the

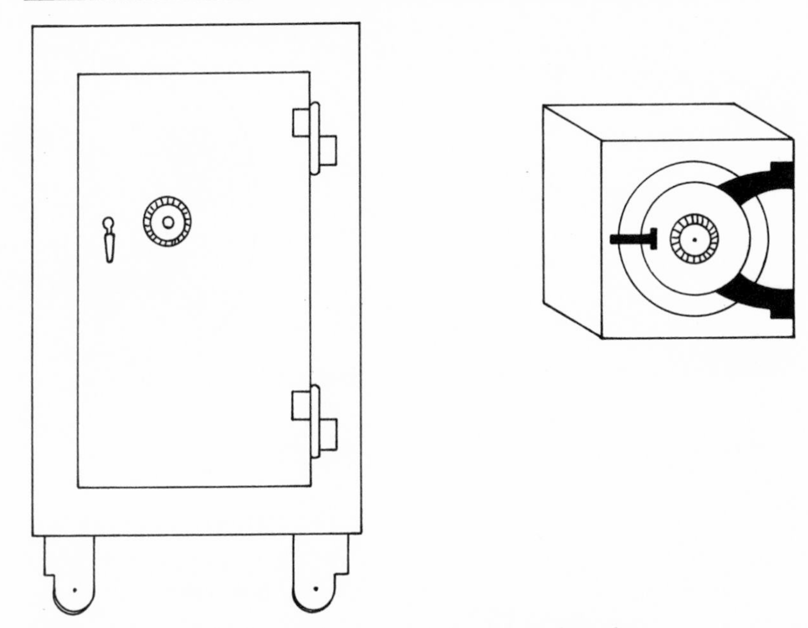

Left Safe, primarily designed to protect documents from fire.

Right Chest, or "money safe," safeguards property from theft.

door is always the thickest part of a chest, this is very unrealistic. Burglars attacking chests almost invariably burn them through the top, using an oxyacetylene torch. If the unit is encased in concrete, the burner chisels away the concrete until the metal top surface is exposed. Understand, then, that the TRTL rating for your chest is based on the material or hardening on the door and front surfaces only. The other surfaces will not resist for as long a period as the label would seem to indicate.

We do not mean to impugn the standards set by UL. We are merely explaining the labeling system; use these classifications as a guide, but be aware what they mean.

In our opinion, the best safe-and-chest-vault units available in this country, particularly for commercial high-risk premises, are the Fortress-Banker-Diamond models made by the John Tann Safe Co. of England. A lighter model, the Stratford, should be considered by anyone interested in safeguarding lesser valuables. The Murray Safe Co. of Brooklyn, New York, is the major U.S. distributor.

For your peace of mind

With the right kind of locks, alarm system, safe or chest, chain bolts, and other hardware, you should no longer feel an oppressive fear of intrusion by burglars. You may be reluctant to take *all* the precautions we suggest, refusing—you might say—to live as a prisoner in your home. You're only kidding yourself. It's better to take much of our advice now than to be driven to improve your security after a shocking invasion of your home. Knowing your burglar's weaknesses and what he fears, you can forever after enjoy a good night's rest.

Index

Numbers in italics indicate illustrations.